8,01/20

Fire: A Very Short Introduction

VERY SHORT INTRODUCTIONS are for anyone wanting a stimulating and accessible way into a new subject. They are written by experts, and have been translated into more than 45 different languages.

The series began in 1995, and now covers a wide variety of topics in every discipline. The VSI library currently contains over 650 volumes—a Very Short Introduction to everything from Psychology and Philosophy of Science to American History and Relativity—and continues to grow in every subject area.

Very Short Introductions available now:

Available soon:

For more information visit our website

www.oup.com/vsi/

Andrew C. Scott

FIRE
A Very Short Introduction

OXFORD
UNIVERSITY PRESS

OXFORD

UNIVERSITY PRESS

Great Clarendon Street, Oxford, OX2 6DP,
United Kingdom

Oxford University Press is a department of the University of Oxford.
It furthers the University's objective of excellence in research, scholarship,
and education by publishing worldwide. Oxford is a registered trade mark of
Oxford University Press in the UK and in certain other countries

First edition published 2020

Impression: 1

Published in the United States of America by Oxford University Press
198 Madison Avenue, New York, NY 10016, United States of America

British Library Cataloguing in Publication Data
Data available

Library of Congress Control Number: 2020930791

ISBN 978-0-19-883003-0

Printed in Great Britain by
Ashford Colour Press Ltd, Gosport, Hampshire

For my grandchildren
The future is in your hands

Contents

Preface and acknowledgements

There will be few of us who will not be frightened by the word 'Fire'. For all of us fire in a building is something we dread and there is scarcely a week goes past that we do not hear of a tragedy involving a building fire. In addition wildfires often reach our television or smartphone screens and an unfolding disaster is documented and discussed. But how much do we understand fire and can we by learning about fire manage to live in a world where there is so much of it? This book has evolved from the development of our understanding about fire, especially over the past fifty years. Many could have written this book but I thank colleagues who have shared their fiery journey over many years, and thank David Bowman, Jennifer Balch, William Bond, Stephen Pyne, Marty Alexander, Stefan Doerr, Chris Roos, Deborah Martin, Jon Keeley, John Moody, Susan Cannon, Rob Gazzard, Susan Page, Juli Pausas, Tom Swetnam, Sally Archibald, Fay Johnson, Cristina Santín, Martin Wooster, Guido van der Werf, Lucian Deaton, Paul Hedley, Margaret Collinson, Claire Belcher, and Ian Glasspool. I thank Lynne Blything for redrawing some of the figures. I would also thank the referees for helpful comments, Latha Menon for her editorial skill, and Jenny Nugee for her help throughout the writing and editorial process.

List of illustrations

Fire

Chapter 1
The elements of fire

Fire is one of those very emotive words with a wide range of connotations, not least in the way it is spoken. Inside a building, an open fire can conjure up heat, light, and comfort, yet an exclamation of 'Fire!' suggests an out of control conflagration that needs to be either immediately extinguished or fled from. The escape routes in public buildings and at public events are usually well explained and signposted. In landscapes with much vegetation, 'fire' might conjure up thoughts of a campfire, cooking, and camaraderie but also danger, if the fire has escaped control, threatening greenery and also nearby settlements. 'Fire' doesn't have to originate with humans: wildfires that cause huge destruction may also arise from lightning strikes and other natural phenonema.

Fire, then, is a potent force, so much so that many consider its control to be an important defining factor of what it means to be human. From the earliest times, fire has been recognized as one of earth's major forces. In Greek myth, fire was a power used only by the gods, until it was stolen by Prometheus and given to mankind. But just as fire can be controlled and tamed it can just as easily be uncontrolled and destructive, so Prometheus has been considered by some as one who strived for scientific knowledge, but who also risked unintended consequences.

Fire was one of the fundamental elements in many ancient civilizations. The ancient Greeks divided the material world into four elements—air, earth, water, and fire—that ranged in their properties from cold to hot and wet to dry, so that air represented hotness and wetness, earth represented coldness and dryness, water represented coldness and wetness, and fire represented hotness and dryness.

Fire has long been used ceremonially in many ways, both secular and religious. A famous secular example, as used today, is the lighting of the Olympic flame. Fire has many meanings in the world religions, and can represent both good and evil. The relationship of fire and evil is most vividly seen in the Western tradition in the fire of hell which punishes sinners, as strikingly described in Dante's Inferno, while fire may also represent purification and escape for the soul in many cultures.

Our ideas about fire have been shaped not only by such deep cultural associations but also by major historical fires that have lingered long in the memory. These may be the destruction of individual buildings, whole towns, or cities as well as wildfires that have devastated large areas of vegetation. But before we look at some of these fires and the questions they raise, we must first consider the nature of fire.

What is fire?

Robert Hooke was one of the earliest to undertake scientific experiments on fire. He describes these in his monumental work *Micrographia*, published in 1665. Hooke was able to determine that the removal of air caused a flame to be extinguished. It was, however, not until the 1770s that oxygen itself was discovered to be the relevant component of air that allowed combustion, by Joseph Priestley (and independently by Carl Wilhelm Scheele), and named by Antoine Lavoisier. Much later, a delightful and accessible description of the nature of fire was given by

Michael Faraday in his popular work 'The chemical history of a candle', published in 1849, based upon his six public lectures on the theme given at the Royal Institution. Even today, we can appreciate his descriptions and observations (Figure 1).

Fire is an exothermic chemical reaction—that is, it produces energy in the form of heat and light—in which a fuel combines rapidly with oxygen (it is rapidly oxidized), leaving a range of reaction products. The fuel is usually a carbon-based compound such as wood, which is made up of cellulose and lignin. Cellulose, the compound that makes up 70 per cent of many plant cell walls and is used to make paper, does not simply react with air at room temperature. But if cellulose is exposed to high temperatures, it

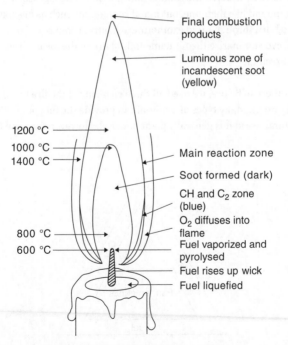

1. **The parts and heating effects of a burning candle flame.**

breaks down into several component gases, including carbon monoxide and methane (CH_4), and methane may react rapidly with oxygen to start a chain reaction.

When methane reacts with oxygen, the two products of combustion are water vapour and carbon dioxide. This is the simplest reaction. Other carbon compounds produce a range of reactive substances that may include carbon monoxide, ammonia, and liquid tars, and some of these also combine to form new compounds such as soot. Water vapour, carbon dioxide, and soot are all elements of smoke, the colour of which depends on the mix of compounds it contains.

We have considered the combustion of a solid, and a gaseous compound like methane, but liquid compounds such as oils may also burn through rapid exothermic oxidative chain reactions. We can summarize the fundamentals of fire in the form of the 'fire triangle'.

For there to be fire, we need all three elements of the fire triangle (Figure 2). Many types of material can provide the fuel but in the natural world it is generally plant material. However, for a fire to

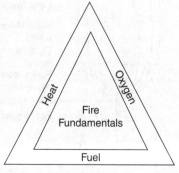

2. **The fire triangle.**

be sustained and propagated there must be a quantity of fuel for the combustion reaction, and the fuel needs to be in a condition for the reaction to proceed. For example, if the plant material, such as wood, is too wet, then much of the heat in the initial reaction will be used to drive off water rather than to burn. The initial heat shock on the fuel will cause the carbon molecules (biopolymers) cellulose ($C_6H_{10}O_5$) and lignin ($C_{31}H_{34}O_{11}$), the main components of wood, to break down to form a range of gases such as carbon dioxide, carbon monoxide, and methane. These gases combine with oxygen in the air to produce an exothermic reaction generating heat and light. The heat produced then causes more fuel to break down and the chain reaction can continue.

In the natural world we tend to think of plants, particularly trees or grasses, being the major fuels but exposed peats, even coals, and oil or gas seeps from the ground may also represent fuel to burn.

In the built environment there is a much larger range of potentially flammable materials. These are not simply wood, as would be expected, but also many oil-derived compounds such as plastics, varnishes, and other oil-based products. Depending on the temperatures reached, even some metals can become available for combustion.

The second side of the fire triangle is oxygen. Our atmosphere today contains 21 per cent oxygen. This is sufficient for fires to burn. However, we know from experiments that below 15 per cent oxygen fires will not sustain themselves. Some recent research suggests that the figure may be nearer 17 per cent. This is why one method of putting out fire is to restrict the supply of oxygen to the burning material. In the built environment this may mean smothering the fire with a fire blanket, with sand, or with carbon dioxide or a foam. We also know from experiments that increasing the amount of oxygen will create a hotter, more vigorous fire.

It is for this reason that no smoking is allowed in hospitals or other places where there are oxygen cylinders in use.

The third side of the fire triangle represents heat. In the natural environment there are three main sources of heat. These are sparks from rolling rock, from volcanic activity, and, most importantly, from lightning strikes. In addition, in some cases spontaneous combustion can occur as heat builds up in fuels as a result of a number of natural processes. Since the evolution of humans we can add human-caused ignitions to the mix. The proportion of lightning-ignited and human-ignited wildfires varies across the globe. In the western USA, for example, depending on the year between 50 and 89 per cent of fires are started by human activity and in the Amazon (and even in England) human activity accounts for more than 99 per cent of fires. Yet in other parts of the world, such as in southern Africa, north-western China, or northern Canada, for example, lightning-ignited wildfires are much more frequent.

There are several forms of lightning, but it is cloud to ground lightning that is the most significant as a source of wildfire ignition. If the lightning hits a plant such as a tree, a short burst of heat will be produced. This may be up to five times the temperature of the surface of the sun. In many cases the energy of this short burst will only be sufficient to vaporize the water in the plant. But if the fuel is dry enough then the heat energy will cause the break-up of the carbon compounds in the plants which when mixed with oxygen in the air will create a chain reaction and burning will begin. So the amount of water in the plant material is key. That's why fires tend to start only during dry periods, when the plant material has a reduced moisture content, and why adding water to a fire is a method used to extinguish it.

Clearly in the natural environment weather plays an important part in creating the conditions for fire, but so does the fuel that is to be burned. Some plants, such as pines and eucalypts, contain natural resins and oils that are particularly volatile and burn easily. One indicator of fire intensity (the amount of heat released) is flame length, as measured vertically from the average flame tip to the middle of the flaming zone at the base of the fire. Even in heather heathlands where the flame length is usually low the occurrence of gorse can create significant flame lengths that can help spread the fire (Figure 3). Gorse (*Ulex europaeus*) is a highly adapted fire species that can cause significant problems even at the edge of a town, the rural–urban interface. Plants that thrive in wet environments are less likely to be subjected to fire but as a consequence they may not have evolved strategies to cope with fire when it does arise. In contrast, vegetation that is found in regions with extended dry periods (such as Mediterranean climates) is much more prone to fire and consequently may have evolved strategies to cope and even thrive in a regime with frequent fires (e.g. chaparral brushfields in southern California).

Extended periods of dry weather may cause the fuel to dry out to critical levels. As moisture content of the fuel is key to the initiation and spread of fire, it represents an important part of what is termed 'fire weather'. A second aspect of the weather that plays a significant role is wind. Wind acts in two ways. It may cause rapid replacement of the oxygen that is key to sustaining the fire, but it also serves to drive the fire. It has recently been shown that the forward rate of forest fire spread is around 10 per cent of the wind speed. Flames may move very quickly and a fire front may move very fast, as fast as or slightly faster than a very fit person can run. In addition, flaming embers may be blown by the wind and fires 'spotted' ahead of the advancing fire front. Spotted fires have been documented to occur upwards of 33 km ahead of the main fire front and may be responsible for creating new fires as well as a mosaic of burned and unburned areas on the landscape.

Fire also behaves differently depending on topography. In mountainous areas hill slopes can produce an updraught effect,

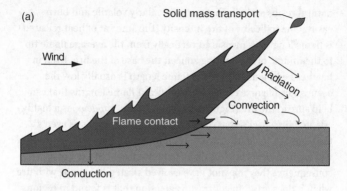

(a)

Solid mass transport

Wind

Radiation

Convection

Flame contact

Conduction

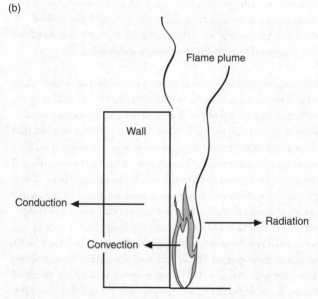

(b)

Fire

Flame plume

Wall

Conduction

Radiation

Convection

3. Differing methods of heat transfer during (a) wildfire and (b) room fire.

creating conditions for a more rapid spread of fire than comparable fuel and weather conditions on flat terrain. The amount and nature of the fuel can also play an important part in how fires spread.

Types of fire

We can divide fuels in the landscape into a number of different categories. The first division is into live and dead fuels. Dead fuels tend to have less moisture than live fuels but in addition dry quicker in warm, rainless weather. Most fires will start in dead fuels lying on the ground surface. These fuels consist of both the surface vegetation in the form of living plants (including herbaceous plants as well as grasses and shrubs) and dead plant litter. The amount and compactness of the fuels may prove very significant in how fast the fire moves and in turn the flame dimensions that are produced.

The second category of fuels is those, mainly living, plants that occur in the crowns of trees, well above the soil surface. Some trees may be dead, in which case they may burn more readily. The most significant crown fuels are from trees, often covering significant areas, that have been killed, for example by beetle or fungal attack.

While these fuels may be significant, a third category known as ladder or bridge fuels provide a connection between surface and crown fuels. These ladder fuels may include tall understorey shrubs, small conifer trees, lichens, hanging moss, dead needles, and branches, or in some cases vine-like plants (lianas) that wrap themselves around the trunks of trees.

The fourth category of fuel includes those materials that lie within or below the soil surface. In some cases this may simply be a humus layer and in other cases it may be a significant peat layer that can catch fire with disastrous consequences.

The distribution of fuel, therefore, gives rise to a number of distinctive types of fire (Figure 4). When lightning strikes most fires start as a surface fire. Depending on the fuel accumulation and the type of fuel, the fire may remain burning on the surface. Such fires may be slow moving, with low flames and low temperature, sometimes only around 400 °C. In many types of vegetation such as grasslands (savannas) or heathlands the fire may remain simply as a surface fire. As these types of fire are small in dimension, they may have minimal effect on the soil and indeed most of the woody plants may not be killed. This is why, for example, a rainstorm following such a surface fire may stimulate a fresh flush of plant growth. In such cases minerals in herbaceous plants may get readily recycled.

If, for example, the vegetation is rooted in an organic rich soil, such as heather on top of peat, then it is possible, when dry enough, for the underlying organic-rich soil or peat to start to burn. In such cases a ground fire may start that tends to smoulder or spread below ground rather than on the surface. In some cases these ground fires may continue to burn long after a surface fire has passed and they can be difficult to extinguish. In some cases the fire may move into the rock layer and if there is coal then this may also be ignited. Underground coal fires may become a significant hazard, burning for many years, and release significant amounts of carbon dioxide into the atmosphere.

If there are significant amounts of fuel on the soil surface, either as accumulated dead fuel or dry living shrubby or herbaceous vegetation, then the surface fire may become more intense. If there are trees present then the fire may spread up the trunk via ladder fuels into the crowns of the trees. Here there may be large quantities of fine fuels that if sufficiently dry and dense may ignite and burn. This will especially be the case in trees with small leaves or needles, such as conifers, although many species of trees may also behave in a similar manner. When the crown of the tree is ignited then the fire may spread through the tree canopy and

**GROUND
FUELS**

FOLIAGE
LIMBS
LOGS
BRUSH
GRASS
DUFF
ROOTS

**AERIAL
FUELS**

FOLIAGE
BRANCHES
SNAGS
MOSS

**ORGANIC
LAYER**

MINERAL
SOIL

4. Types of fuel and fire: (a) Surface fire; (b) Crown fire; (c) Ground fire.

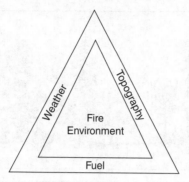

5. Fire environment triangle.

become a crown fire. Such fires may become very intense and the moving fire front will dry out the vegetation in front so that it catches fire more easily, allowing the fire to continue its forward propagation. Crown fires may move very rapidly, especially if there is significant wind and it is possible for the crown fire and surface fire to separate and move at different speeds. The combination of topography (fires spread faster up a slope as heat rises providing an updraft and pre-heats the vegetation in front) and wind may help drive the fire, sucking in oxygen and sometimes creating what is termed a fire-storm.

Even in cases of intense crown fires only a small amount of the total available fuel is consumed. It is possible that large swathes of dead vegetation may remain that have the potential for fuelling another fire soon afterwards. We can summarize these aspects using what we can term the fire environment triangle (Figure 5).

Fires that changed the world

The fires with which most of us are familiar will not necessarily be ones that we have experienced but ones that have grabbed our attention for one reason or another (Tables 1(a) and 1(b)). It is quite possible that our first encounter with a significant fire is one

Table 1(a). Major landscape fires

Date	Place	Landscape fire	Size* (hectares)	Deaths
1910	USA	Year of the fires across USA	1,200,000	87
1939	Australia	Black Friday Bush Fires	2,000,000	71
1982/3	Indonesia	Large area of Kalimantan (Borneo) burns	550,000	
1988	USA	Yellowstone fires, Montana	321,270	
2002	USA	Hayman Fire, Colorado	56,750	5
2003	Portugal	Destroys 10% of Portuguese forests	215,000	
2005	Spain	Guadalajara Fire	13,000	11
2009	Australia	Black Saturday Fires	1,100,000	180
2010	Russia	Siberian fires	300,000	N/A
2011	USA	Los Conchas Fire	63,250	
2011	England	Swinley Forest Fire	90	
2013	USA	Rim Fire, Northern California	102,520	
2013	USA	Yarnell Hill Fire, Phoenix, Arizona	530	19
2016	Canada	Fort McMurray Fire, Alberta, Canada	593,670	

The elements of fire

(*continued*)

Table 1(a). Continued

Date	Place	Landscape fire	Size* (hectares)	Deaths
2017	Canada	British Columbia Fires	1,354,284	
2017	Portugal	Fires in June and October (7,580 fires)	54,000	111
2017	Chile	2,977 fires across Chile	505,850	
2017	USA	Thomas Fire, California	114,078	
2018	Greece	Attica fires		102
2018	England	Saddleworth Moor Fire, England	720	
2018	Canada	British Columbia	1,298,450	
2018	USA	Fires in California including the Camp Fire	60,000	85
2019	New Zealand	Nelson Fire, South Island, New Zealand	2,300	
2019–20	Australia	Fires across Australia, especially eastern Australia	18,600,000	34

* 1 hectare = 0.01 sq. km, e.g. 60,000 ha = 600 sq. km.

we may have read about when young. A good example would be the legend of the burning of the ancient city of Troy, carried out by the Greeks using the trick of the Trojan Horse, which allowed the attackers to enter and burn the city to the ground. This reflects the deliberate use of fire in warfare, which has occurred throughout history. A 20th-century example was the burning of Dresden in Germany by aerial bombardment by UK and US forces in the Second World War, and incendiary devices continue to play a role in modern wars. Other famous examples began accidentally. The Great Fire of London of 1666, recorded by the diarist Samuel Pepys, thought to have begun in a baker's shop, led to the rebuilding of London, with important contributions from Sir Christopher Wren and Robert Hooke. Other fires that ravaged towns or cities include Tenmel in Japan in 1788, Moscow in 1812, Chicago in 1871, and San Francisco in 1906. Each of these tragedies led to an improvement in city and building design as well as the development of fire prevention measures and fire suppression within the urban environment. Some lessons have been learned with improvements in fire safety and, as such, large, accidental catastrophic fires are much less likely.

Single or even multiple building fires, even when accidental, can provide an impetus for a reassessment of fire prevention and suppression in an urban or built environment. A recent example is the tragic Grenfell Tower Block Fire in London in 2017, where the problem with the external cladding of the high-rise building was a significant factor in the spread of the fire that ultimately led to seventy-two deaths.

Just as in urban settings major fires in wildland areas may change both the view of fire and behaviour towards such fires, and sometimes this may have unintended consequences. As most wildland fires occur away from population centres, they may not have come to the attention of the public consciousness. One of the first such events was the large number of fires that took place across the western United States of America in 1910. This led to a public outcry for something to be done. A major campaign of fire suppression ensued in the 1940s with the iconic Smokey the Bear leading the way. Even relatively recently forest fires were seen as a tragedy, as illustrated in a postage stamp from the USA. However, the complexities of fire were not known at that time and in some

Table 1(b). Major city and building fires

Date	Place	Comments	Deaths
586 BCE	Jerusalem	Burning of first temple	
48 BCE	Alexandria	Burning of the Library	
64	Rome	Great fire set by the Emperor Nero destroys much of Rome	
1577	Venice	Burning of Doge's Palace	
1666	England	Great Fire of London spreads from bakery shop across London	
1760	Boston, USA	Great fire of Boston	
1788	Japan	Great fire of Tenmel	150
1812	Russia	Fire of Moscow	
1871	USA	Fire of Chicago	250
1881	Austria	Ringtheater fire, Vienna	384
1899	USA	Windsor Hotel fire, Manhattan	45
1906	USA	Great fire of San Francisco	
1923	Japan	Great fire of Tokyo	100,000
1938	China	Changsha fire	3,000
1945	Germany	Fire bombing of Dresden	30,000
1987	England	King's Cross, Underground fire, London	31
1988	USA	First Interstate Tower fire, Los Angeles	1
1992	England	Windsor Castle fire	
2001	USA	11 September attack on World Trade Center, New York	2,600
2005	England	Buncefield oil storage fire	
2010	China	Shanghai Apartment fire	53

Fire

2012	Pakistan	Karachi Garment Factory fire	312
2015	UAE	Torch Tower Fire, Dubai	
2017	England	Grenfell Tower Block fire, London	72
2019	France	Notre-Dame de Paris	

areas putting out the fire (where there had been regular surface fires previously keeping surface fuel loads low) led to the build-up of fuel so that when a fire eventually started it quickly became an intense surface fire that resulted in a large crown fire with more devastating effects.

We might consider the large fires that occurred across the island of Borneo in 1982/3 as being an event that caught the general public's attention and was highlighted by the World Wildlife Fund, but which led to a greater understanding of the problem of wildland fires caused directly by human impact. In this part of the world there are extensive peatlands upon which tropical rainforest grows. The trees in the rainforest had been a major source of revenue for loggers so in many cases the peatlands had been drained for logging to take place. Fires starting in the resulting dry fuel can become quite large and destructive, not only producing large amounts of smoke that has been shown to be hazardous to human health but also releasing large quantities of carbon dioxide into the atmosphere from the burning of the peat. In recent years the problem has been exacerbated by the draining of peatlands to allow the planting of oil palms, again leading to major fires with deleterious consequences.

In many fields there are defining moments or discoveries. One such revelation in relation to wildfires came in 1988 following major fires in the Greater Yellowstone area of Wyoming and Montana, USA. Large areas were burned by over forty separate fires, and the Yellowstone National Park was closed to visitors.

These fires created headlines across the globe. Although some of the fires were started by humans, in fact most were started by lightning strikes. It has been claimed that the fires became large as a result of the long-term consequences of attempted fire exclusion. It may have been a more complex matter but the discussion led to a reassessment of both the policy of fire suppression and prescribed burning in the USA. What we now believe is that different vegetation types may have different fire regimes and a 'one size fits all' policy for wildfire may not be appropriate. Another consequence of the 1988 Yellowstone fires was the identification of the scale of post-fire erosion and deposition of large amounts of sediment following the fire and after major rainstorms.

The large 2002 Hayman Fire, near Denver, Colorado, provided an opportunity for extensive studies to be undertaken both during and after the fire. Satellite imagery was effectively used both to track the fire and also help in the post-fire recovery assessment.

Outside of the United States the 2009 Black Saturday fires in Australia across the state of Victoria have also had important ramifications. The death total (173 individuals) represented the largest peacetime disaster in the country's history. The initial reaction of politicians was all about blaming the person who started the fire. At that stage there was little discussion about drought and the nature of flammable vegetation and also that communities had built into this flammable landscape. However, the outcome led to a high level report by the Royal Commission that interviewed not only those involved on the ground but also several fire experts. This led to discussion of whether we should build at all in flammable landscapes and, if we do, how we can make our homes and population centres more resilient to fire.

The 2016 Fort McMurray Fire in Alberta, Canada, was headline news around the world, not only because of its scale but also due to how rapidly the fire spread through the community, threatening the escape routes of the population: fortunately, no lives were lost. Could more have been done, given our knowledge of the frequency with which such coniferous forests burn?

Another key example that has rung alarm bells was the June and October 2017 fires in Portugal that swept through *Eucalyptus* (eucalypts) and pine forest plantations. A hundred and eleven people lost their lives. Much of Portugal can experience natural wildfire. After the Second World War large eucalypt plantations were established but the dangers of this action were not realized until many years later. Eucalypt forests tend to be very flammable fuel complexes and if there are not regular planned under burnings to reduce fuel loads, when a wildfire gets hold it can be catastrophic.

Our last example is the 2017/18 fires in California near Santa Monica known as the Skirball Fire. Several extensive fires spread across the area from Los Angeles to Santa Barbara in December 2017, including the large Thomas Fire between Ventura and Santa Barbara. But it was the Skirball Fire that received significant media attention as it closed the road north of Los Angeles and burned the houses of many famous people. In January 2018 following the fires there were major rainstorms across the burned areas that created significant post-fire erosion and mudslides. This was one of the first times post-fire erosion was brought into the public domain, as it affected a well-known celebrity who was filmed wading through mud in her garden.

Major fires can lead to disasters, but they can also be instructive, driving research, changing attitudes, and helping develop new fire policies.

Consequences of fire suppression

In an urban context there is no dispute that fire suppression is necessary. In a wildland context, though, there is not only debate concerning when to extinguish a fire but methods for doing so can be quite diverse.

In a wildland fire situation fire suppression would seem to be obvious. Perhaps this is because of the success of campaigns such as Smokey the Bear in the USA and similar fire prevention efforts in a number of other countries. Clearly, images on the television of raging wildfires engender fear and an urge to put out wildfires. This is particularly the case when homes and people are threatened but even when this is not the case the urge to do so persists. We now know that immediately attempting to suppress every fire is not the answer and we need to discuss when and where wildfires are suppressed.

However, our ability to suppress wildfire in the past fifty years has developed considerably. We should consider three main methods of suppression: to remove the fuels ahead of the combustion zone; to reduce the temperature of the burning fuels; or to smother the fire to exclude oxygen (the approach used by fire beaters in moorland wildfires). Quite naturally we tend to think of putting out a fire using water via a fire vehicle. The type of fire vehicle used to put out wildfires may be quite different from that used in an urban setting especially as water may not be easily replenished. In remote areas water may be moved and dropped by air, either using a helicopter which can access water from lakes, reservoirs, or even the sea, or using larger aircraft that can pick up and transport water over large distances and drop over designated spots rather than on the fire itself or in advance of a spreading fire front. Such a technique is particularly useful in putting out small fires that have been started by embers (so-called spot fires) some distance from the main fire front, but only if the fire is not of high intensity or moving very fast. The water may also be treated with additives that help to extinguish fire and prevent re-ignition.

One of the key ways to prevent the movement of fire that is particularly effective with surface fires is that of fuel clearance in cutting a fire-break. This can be back-breaking work for large fire crews but may be most effective in protecting property. It is particularly important when fire fighting to have a good understanding of fire behaviour, not only considering the speed of fire advance but also the impact of any changing vegetation, topography, and wind speed or direction. Even the most experienced fire fighters can be caught out by the range of factors affecting fire spread, as with the death of nineteen City of Prescott fire fighters during the 2013 Yarnell Hill Fire near Phoenix, Arizona, in the USA.

The expression 'fighting fire with fire' is well known but not always understood in the context of fire suppression. We have seen how fuel reduction may reduce the impact of an intense wildfire so that a surface fire may be easily put out and not develop into a crown fire. A surface fire may be started and controlled, either as a back fire behind a fast-moving fire front, so that a change in wind direction may cause a fire to cease, or else ahead of a fire front, so that the reduction in fuel will make the extinguishing of the fire much easier.

In spite of all these efforts, ultimately it is often a change in the weather that is the final extinguisher of a major fire.

Fire at the wildland–urban interface

The wildland–urban interface or WUI (pronounced 'woo-ee') is a relatively new term but an increasingly useful one where wild areas that may experience a wildfire are being encroached on or invaded by population centres. (In some countries, such as England, the term rural–urban interface may be more appropriate but we will consider both together.) This is a critical boundary or zone as in many cases it divides quite different approaches to fire. Outside the town or city fire may be a regular occurrence, not only familiar to those who live in such areas but in some cases used by people either in land transformation or in agriculture. However, within an urban context fire has to be contained as uncontrolled fire puts both buildings and people at risk. This interface can therefore become a problem between two worlds, one with fire and one without. And humankind is increasingly moving from urban centres to wildland environments, some of which are very fire prone. As we shall see, this raises pressing issues about how to protect human settlements in flammable environments, which means understanding how fire behaves in different landscapes and types of vegetation in a range of climatic regimes. But long before human settlements, or the evolution of humans, fire was a force that altered the Earth.

Chapter 2
The deep history of fire

Identifying past fire events

There are many features of fire systems in vegetation that cannot be assessed without recourse to fire history. We can view fire history on a number of different scales, from the length of a human lifetime to a geological scale of millions of years. Each scale will, however, require a different approach to the gathering of data. To establish the characteristics of fire regimes we need historical data, to learn about, for example, fire return intervals (FRI—the time between fires in a defined area, usually in a single vegetation type) that will aid current policies about wildfire. Today, we have the advantage of recorded information either by individuals or organizations including satellite information. While this is giving us a better global view of fire, we still need a much better perspective of fire through time, especially if we are able to assess the impact that humans have had on natural fire systems and to get to grips with the impact on wildfire of climate change. Apart from written or oral records for recent history, we can access data about fire stored in nature itself.

One approach is to use the growth rings within the wood of trees. We can date a tree from growth rings and obtain a considerable amount of climate data in addition. But growth rings can also tell us about fires of the past. As we have already seen, surface fires are common in many forested areas and can pass through without killing the trees. The trees may, however, be scorched or burnt in one area, giving rise to a fire scar that may be healed over. When the tree is felled, we can then establish not only that a fire has occurred in the past, but also the year in which the fire occurred. In trees with a long run of tree rings fires may have occurred many times and this pattern of scars provides information on the frequency of fire—the FRI.

If trees from a wide area can be studied then this may provide data on the size of a particular fire in the past. If it proves possible to obtain fire data from trees across a whole region then this can be coupled with information on climate and allow an interpretation of how fire frequency is affected by climate change, as well as seeing the effects of major fluctuations caused by the phenomena of El Niño and La Niña, which are the warm and cold phases respectively of a large-scale atmosphere and ocean coupling cycle called the El Niño Southern Oscillation (ENSO). During an El Niño phase, parts of the western Pacific become particularly dry. In some places, such as the Sequoia National Park in northern California, it has been possible to use tree rings to obtain a fire record that goes back nearly 3,000 years.

Obtaining tree ring/fire scar data is limited to forest settings, where trees can be sampled. Fires, when burning, emit smoke into the atmosphere that comprises not only water vapour but complex chemicals such as polycyclic aromatic hydrocarbons (PAHs), gases such as carbon dioxide, nitrous oxides, ammonia, recombined organic carbon in the form of soot, and small charred particles (charcoal, Box 1). While many of these compounds, such as soot, PAHs, and ammonia, have been used to interpret fire events, they have proved controversial, especially when interpreting the size of fires. Small charcoal fragments (termed micro- and meso-charcoal) have more frequently been used in the interpretation of fires that occurred over the past million years or so (Figure 6).

human-created		natural human-influenced			natural
Increasing change in human-nature fire interaction		Agricultural fire	Hominin-fire interaction, megafauna, fire and hunting		First plants, first fire
Modern industrial fire	Industrial fire	Human-fire megafaunal	interaction, extinction	Evolution of grass-fire cycle	

0	1 yr	10 yr	100 yr	1 kyr	10 kyr	100 kyr	1 Myr	10 Myr	100 Myr	500 Myr

6. Types of fire seen through time.

Such studies have proved most effective in interpreting fires over the past 70,000 years, as rock sequences can be readily carbon dated and this data on fire return intervals and climate change can be analysed. Charcoal (generally less than 1 mm in size and often counted in categories greater and less than 100 μm) can readily be deposited in lake and peat sediments where there may be a long history of sedimentation and where age profiles of the sedimentary sequence may more easily be obtained. Researchers have been encouraged to pool their data to create a global charcoal database that can be used to study fire patterns on both a regional and global scale.

The data collected by a number of different teams of scientists need to be standardized and data on charcoal flux may be calculated and departures from a background norm can be used to identify significant fire events. The record becomes more common over the past 2,000 years and has proved a powerful tool in the interpretation of fire regimes in relation to climate change and in relation to human activity, although not all changes are picked up using this method.

Charcoal records are obtained and interpreted rather differently in the Quaternary and Holocene (the last 2.5 million years) compared to those from earlier times. Small charcoal particles can easily be identified by their black colour and their often lath-like

25

Box 1. The formation of charcoal

When wood is burned there are distinct zones of combustion, charring, and pyrolysis (that is, heating in the absence of oxygen, which causes thermal alteration and decomposition) (Figure 7). This is important as it means that any branch or tree trunk, or indeed any piece of wood, burns in a similar manner. We are able to see this if we look at a piece of wood that has not been fully combusted. We have seen how a high-temperature pulse is needed to initiate the process (although in the rare case of self-ignition the temperature rise may be more gradual). At 20 to 110 °C the wood absorbs heat as it is dried giving off moisture vapour (steam). The temperature remains at or slightly above 100 °C until the wood is 'bone dry'. At 100 to 270 °C the final traces of water are given off and the wood starts to decompose, giving off some carbon monoxide, carbon dioxide, acetic acid, and methanol. Heat is absorbed. At 270 to 290 °C, this is the point at which exothermic decomposition of the wood starts. At this point heat is produced and breakdown continues spontaneously, providing the wood is not cooled below this decomposition temperature. Mixed gases and vapours continue to be given off, together with some tar. Some of this tar may move further down into the charcoal structure and be precipitated there. This may result in the formation of glassy carbon that is often found by archaeologists, and that has been shown to be a result not of very high temperature but of the precipitation of compounds in the charcoal residue. At 290 to 400 °C, as breakdown of the wood structure continues, the vapours given off comprise the combustible gases carbon monoxide, hydrogen, and methane, together with carbon dioxide gas and the condensable vapours: water, acetic acid, methanol, acetone, etc., and tars which begin to predominate as the temperature rises. At 400 °C the transformation of the wood to charcoal is practically complete.

The charcoal at this temperature still contains appreciable amounts of tar, perhaps 30 per cent by weight trapped in the structure. This 'soft burned charcoal' needs further heating to drive off more tar and raise the fixed carbon content of the charcoal to about 75 per cent. To drive off this tar, the charcoal is subject to further heat inputs to raise its temperature to about 500 °C, thus completing the carbonization stage.

While this describes the formation of wood charcoal, a range of plant tissues follows a similar pattern. Despite this alteration of the cell walls, the anatomy of the plant is still preserved (Figure 7). However, mass is lost during this process so the plant drops in weight.

In addition, shrinkage of the various organs can occur. The carbon content of the plant tissues increases and the carbon domains within the cell wall become ordered. The consequence is to make the plant tissues more resistant to decay. This feature is the rationale behind the production of biochar as a mechanism for CO_2 sequestration. The solid material that remains from this pyrolysis process is known as charcoal and itself may be consumed if the combustion process is not halted.

shape on the glass slides of organic residues obtained from rocks showing ancient pollen and spores, known as palynological slides, that have traditionally been used to identify vegetation and its changes in the rock record. This data has been used to interpret fire occurrence, and has led to the idea of constructing a global charcoal database. But while this approach works for recent deposits, more ancient records become more difficult to interpret. The first problem is related to the identification of small charcoal particles on the palynological slides. As organic material is buried, the colour of the organic particles changes because of increased heat experienced through the burial process (it gets hotter as you

(b)

Heat from external source

Original surface

Fissured
Zone

Unfissured
Zone

Residual
char

Pyrolysis
Zone

Unpyrolysed
Wood

← *Direction of grain in wood* →

Solid arrows indicate probable directions
of movement of volatile products

(c)

7. **How charcoal forms.** (a) Charred log showing cracking; (b) Section
of charred wood showing charring zones and movement of volatiles;
(c) Cut tree after wildfire showing only outer zone is charred.

drill down through sedimentary layers) from yellow to brown to black—a process known as maturation or coalification. This means that distinguishing black charcoal particles from dark coalified particles becomes more difficult, although oxidative acids may play a useful role by preferentially changing the colour of the coalified particles from black or brown back to yellow. In addition, absolute dating becomes more difficult in rocks older than 70,000 years as carbon isotopic dating is not viable and other methods lack the precision to obtain yearly, decadal, century, or even millennial accuracy.

Another approach to interpreting fires in the past is to examine macroscopic charcoal (>180 µm but usually over 1 mm in length) that can be extracted from sediment or rock samples. Charcoal fragments in rocks may reach dimensions of centimetre cubes and even larger. These will predominantly have been buried near the fire site (such as in peats) or moved, usually by water, to a depositional site that could have been a river, lake, or even the sea. Experiments on charcoal settling rates and water transport have shown unexpectedly that larger charcoal particles can travel greater distances than smaller particles, as they take longer to sink in the water column.

The study of larger charcoal fragments has one distinct advantage (Figure 8). Charcoal may preserve the anatomical details of the plant being burned, so charcoals may provide evidence not only that there was a fire but also what vegetation was burned.

Surprisingly there have been few studies on macroscopic charcoal assemblages associated with microscopic charcoals from the more recent past. However, it is the data from macroscopic charcoal that have opened up the deep time history of fire although the interpretation of fire frequency can be much more difficult. For the use of charcoal to interpret fire history in older rock sequences it is necessary to be able to recognize such fragments in the rocks. Until now Pre-Quaternary charcoals have been studied in detail in

(a)

(b)

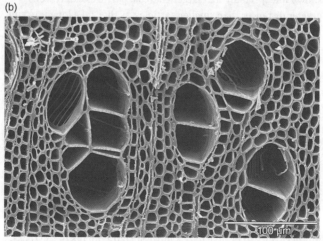

8. Modern and fossil charcoal. Light and scanning electron microscope (sem) images. (a) Fossil wood charcoal from the Jurassic rocks of Yorkshire (scale 1 cm); (b) sem of beech wood charcoal (scale 100 μm).

a limited number of regions and time periods and our understanding of ancient fire systems is relatively recent, developing only over the past forty years or so.

Fire and the evolution of vegetation

It goes without saying that the history of wildfire is intimately tied up with the evolution of plant life on land. Vegetation provides the fuel to burn but the nature and quantity of that fuel has changed through the course of geological history since plants first appeared on land.

The earliest land plants evolved somewhere between 450 and 420 million years ago, during the Silurian Period of the Geological Timescale. However, although there may have been small patches of algae, mosses, and liverworts for a long time, we tend to think of the greening of the landscape as occurring from the evolution of vascular land plants (that is plants with specialized tissues for the transport of water and the products of photosynthesis). These plants had a number of characteristics that allowed them to live on land, including a cuticular covering with water/gas exchange pores (stomata). These green plants absorbed carbon dioxide from the atmosphere and water from the soil, and used energy from the Sun in the chemical process of photosynthesis to transform the water and gas into carbohydrate molecules such as sugars and other chemicals that allowed the plants to build their organic skeleton. In the process oxygen was released into the atmosphere as a by-product. The process of photosynthesis can be summarized in general terms in the following chemical equation:

$$6CO_2 + 6H_2O + \text{solar energy} \rightarrow CH_{12}O_2 + 6O_2$$

As we shall see, this equation has great significance in our discussion of fire.

The second feature of these early vascular land plants is that they produced vascular tissue that comprises several cell types such as the water-conducting elements known as tracheids (xylem) that are made from cellulose impregnated with the strengthening chemical lignin, and phloem, which transports food for the plant.

Thirdly, the plants developed a reproductive strategy that involved producing spores. These spores were shed by the plant (known as the sporophyte) onto the damp soil surface. Here they developed into small predominantly underground gametophytes, which produced the male and female sex organs. Sperm from the male gametophyte swam in the damp soil to reach and fertilize the female gametophyte and a new generation of the plant, the sporophyte, was produced. This type of reproduction can be seen in ferns today.

The earliest plants from the late Silurian Period, around 420 million years ago, included the diminutive form known as *Cooksonia*. Indeed many of the earliest land plants were only a few centimetres tall, so they would not provide sufficient fuel for a large wildfire. In spite of that our earliest charcoalified plants that demonstrate the occurrence of fire on Earth came from this Period. Such fires would have been small and very localized. These early plants were restricted in their habitats to being near water and were patchy in their occurrence.

Through the subsequent period of geological time, the Devonian (419–358 million years ago), plants evolved a number of new strategies that allowed them to spread into new environments and increased potential fuel loads. Plants evolved a range of growth habits that helped in their spread. One of these was to reproduce by a mechanism known as clonal growth (where the plant reproduces vegetatively from a single individual). In such cases the plants may be connected underground. But all early land plants were restricted by their method of growth. They were only able to grow by primary growth via their growing tip (meristem), so growing tall was a significant problem. Some plant groups of the Devonian overcame this by developing a method of secondary growth: new cells are produced not at the growing tip alone but also around the circumference of a stem, in a layer between xylem and phloem known as the cambium. The cambial cells produce secondary xylem and phloem cells. These secondary xylem cells (wood) are chemically composed of around 70 per cent cellulose and 30 per cent lignin, so they provide significant strength. The ability for the plant to increase in girth allows the plant to become taller and develop an arborescent (or tree) habit. This also helps the plant live longer, and a larger potential fuel load is produced.

However, growing taller produces other effects too, not least the need to increase photosynthetic activity. Another structure evolved during the Devonian that solved this problem: the leaf. The increased area available via leaves to capture sunlight and photosynthesize not only helped some plants in their growth strategy but also created some shade on the soil surface, and this led to a further diversification of plants adapted to an ever-increasing range of habitats.

These early plants, even trees, still reproduced by spores so that the plants needed to have a ready source of water. It was the evolution of the seed habit in the late Devonian, around 370 million years ago, that finally allowed plants to spread into much drier habitats.

The spores that produce male and female gametophytes became differentiated, with those that evolve into female gametophytes growing bigger, into megaspores encased in ovules, and becoming retained on the plants and only released when they were fertilized to form a seed. The development of seeds allowed the plant to provide an increased food reserve for the new plant and to release it from the need to grow in damp soil. The seed habit allowed plants to grow and thrive in much drier habitats.

By the mid to late Devonian, as plants grew larger, providing more fuel, and spread into drier environments, larger and more frequent wildfires must have become possible. Yet this pattern is not seen in the record of fossil charcoal. As we have seen there are three parts of the fire triangle—fuel, heat, and oxygen. But in deep time, we need to consider a slightly different triangle (Figure 9).

One side of the deep time triangle represents the evolution of plants or fuel; the second represents the evolution of the atmosphere, especially oxygen in the atmosphere; and the third represents climate. If we had a significant build-up of fuel by the mid to late Devonian and the plants were living in drier climates then why do we not see more indications of fire in the rock record? The answer almost certainly lies in the oxygen content of the atmosphere.

Fire and the atmosphere

For sustained fire we need a minimum of 15 per cent oxygen in the atmosphere, though recent research suggests a more realistic figure may be 17 per cent. Today the figure is 21 per cent. If we have a build-up of fuel on the landscape with dry conditions then the absence of fire may be the result of low atmospheric oxygen.

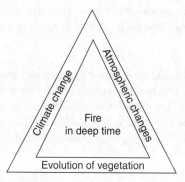

9. **Deep time fire triangle.**

The occurrence of fire in the late Silurian, 420 million years ago, suggests that oxygen levels were at least 17 per cent. The subsequent absence of fires for much of the period 400–360 million years ago could be a result of the atmospheric oxygen levels falling below this critical figure. The calculation of atmospheric oxygen has come traditionally from geochemical models that have used a range of data to predict oxygen through time. Some of these models suggest that there was a fall in oxygen levels through the Devonian and this may have been the reason for such limited or no fire activity. These models also suggest that oxygen levels rose again in the very latest Devonian and into the Carboniferous Period (359–340 million years ago) and we also see an increase in charcoal in rocks of this age, providing evidence of wildfire activity.

A number of geochemical models also indicate that from around 350–250 million years ago oxygen levels were significantly above the modern level of 21 per cent. Some models have suggested levels between 30 and 35 per cent oxygen during this period. The implications of this are significant. Experiments have shown that increasing oxygen content above 21 per cent significantly increases the possibility of fire. Increasing oxygen content allows wetter plants to burn. Some experiments have indicated that in an atmosphere of between 30 and 35 per cent oxygen even very wet plants could burn, and so fires would be difficult to extinguish. Clearly such levels of oxygen are unsustainable as the vegetation would always be burning. New models that have included fire feedbacks in their calculations still, however, indicate oxygen levels higher than that of the present throughout this time period.

There are no obvious proxies for atmospheric oxygen but it has been suggested that the charcoal fossil record might provide some clue as to changing levels. This idea is based on the amount of charcoal that is found in ancient peats that, through time, have been transformed into coal. We know that peats were formed in wet conditions not prone to fire, so that in modern peats there is less than 5 per cent charcoal present globally. For oxygen levels below 21 per cent there would be even less charcoal present, and none at all at 17 per cent oxygen. As oxygen levels rise up to 30 per cent the very wet plants may burn and charcoal levels can be expected to increase. Data from charcoal content in coal indicates that in the late Paleozoic Era—the Carboniferous and Permian Periods—many coals have over 30 per cent charcoal and some have over 70 per cent. Such data can be used to calculate atmospheric oxygen levels and the charcoal record shows, along with the traditional geochemical models, that the late Paleozoic was a period of high oxygen and also high wildfire activity.

Both approaches also show that oxygen levels varied considerably over the following Mesozoic Era (encompassing the Triassic, Jurassic, and Cretaceous Periods, 250–66 million years ago) but the mid to late Cretaceous (120–66 million years ago) was also a time when oxygen levels were higher than today. All data further suggest that the modern levels of oxygen were stabilized at around 21 per cent around 40 million years ago.

Fire and climate

In the world today climate plays a fundamental role not only in the occurrence of different vegetation types but also in the occurrence of wildfire. There are several aspects of climate that need to be considered. The first is temperature. As temperatures rise, fuels may dry and make fires more likely. However, rainfall is also important: if it is wet, then in spite of temperatures being high fire will not take hold, even if started by lightning strike. More subtly the length of dry and wet periods may play an important role. If there are wet winters and springs yet dry summers this may lead to fire activity but if it is wet all year long then fire would be unlikely (rainfall limited). In areas of prolonged dryness plant growth is much reduced, and as a consequence, fuel build-up would be less as fire that could more easily start because of the dry condition of the fuels would not develop because of the lack of fuel to burn (fuel limited).

The differences in both rainfall and temperature not only play an important role in the types of vegetation that may be found but also in the occurrence of fire in these different environments. We might not expect much natural wildfire in tropical rainforests as a whole as while there is a significant amount of potential fuel the rainfall does not allow it to dry naturally and hence be available for fire. The intervention of humans may change this. This is one reason for the outrage generated by the deliberate starting of fires across the Amazon rainforest in 2019. Mediterranean-type climates with long hot summers make for very fiery regimes but even in cooler areas such as in boreal forests periods of dryness may lead to extensive fires. It is a surprise to many that Alaska experiences large numbers of wildfires.

The distribution of fire on Earth has only been fully understood with the development of satellite monitoring. It has allowed us to understand more about natural flammable vegetation biomes as opposed to those in which fire is unnatural but may potentially be modified by human activity.

Low and high fire worlds

The realization that atmospheric oxygen levels play an important role in the occurrence and amount of fire through time has led to the idea of low and high fire worlds. We may consider that the Devonian Period was a low fire world, initially because of limited fuel but later because of reduced atmospheric oxygen concentration. In contrast, the Late Paleozoic was a high fire world even though climate changed considerably through that time, and it was predominantly an icehouse world with polar ice. But while the history of fire through these early times provides many interesting insights into how fire and the biosphere evolved, it is the history of fire through the subsequent periods that has most relevance to our current understanding of wildfire.

The vegetation of the Mesozoic, in particular the Triassic and Jurassic (250–145 million years ago), was dominated by seed plants and the oxygen levels varied significantly over the time period but did not fall below the 17 per cent level. By far the most important changes took place on Earth, however, during the Cretaceous Period (140–66 million years ago).

One of the most important changes in vegetation occurred during the Cretaceous, with the evolution of flowering plants (angiosperms). These plants developed new ways of reproduction via flowers and pollination that were successful both in wind but also most especially through insect pollination. During the early part of the Cretaceous (140–100 million years ago) land floras were dominated by a range of seed-bearing plants including conifers, cycads, and cycad-like plants known as the Bennettitales. Ground cover was dominated by ferns and horsetails, both spore-bearing plants. The atmospheric oxygen levels rose through the Cretaceous and by around 120 million years ago many types of vegetation burned, both conifer and fern-dominated types, extensively during this high fire world.

This, however, was when the flowering plants first evolved. The changes in fuel structure may have significantly altered fire regimes. Many early flowers were weedy plants that thrived in disturbed environments. So frequent fires would have aided the spread of these early flowers and many flower fossils found from this period are preserved as charcoal (Figure 10).

Our understanding of this high fire world and its impact on the evolution of life has been considerably enhanced by molecular studies looking at the DNA of modern plants. For example, both pines (conifers) and Proteaceous plants have been predicted using molecular clock techniques to have evolved their ability to survive

10. Scanning electron micrograph of fossil charcoalified Cretaceous flower.

fire during the Cretaceous, and recent fossil finds have tended to support this hypothesis.

Fire is likely to have played a significant role in landscapes that were dominated by the dinosaurs. The extinction of the dinosaurs and many other animals and plants in a mass extinction some 66 million years ago has been linked to a large asteroid impact. The impact is quite well established, and left the massive Chicxulub crater in Mexico, and many accept that it played a significant role in the mass extinction, although large-scale volcanism and climate changes are also likely to have been major causes. The suggestion of a global wildfire following the impact has, however, received much less support. Research suggests that there is insufficient evidence for a global wildfire and also experiments and a general understanding of wildfire dynamics suggest that such a global wildfire was unlikely, even in the high fire world of the Cretaceous.

After the Cretaceous, changes in climate, with increased rainfall together with the evolution of tropical rainforests around 40 million years ago, suppressed fire activity. The move from a relatively high global temperature during the Eocene Epoch (46–34 million years ago), to a much cooler world (going from a greenhouse to an icehouse world) from 20 million years ago until the present, shaped the development of modern vegetation and fire regimes. However, a significant change took place around 7 million years ago, during the Miocene Epoch, with the spread of savanna grasslands, especially in Africa.

Grasses first evolved during the early part (between 66 and 30 million years ago) of the Cenozoic Era (66–2.5 million years ago) but these had a traditional biogeochemical pathway for photosynthesis known as the C3 pathway. However, during drier intervals some grasses developed an efficient new pathway known as C4. This helped grasses thrive and spread in drier climates and soils. The rapid growth of grasses in drier habitats also provided a significant surface fuel load, giving rise to a grass-fire cycle, which itself aided the spread of such grasses and resulted in the development of savannas such as are seen in Africa today. While fire may burn off dead dry plants the roots of the grasses are not killed. If fire occurs regularly (i.e. every ten years or so) then competing larger plants that may grow into shrubs or trees are killed. Fire, then, is important for the maintenance of large tracts of savanna in many parts of the world. We may then consider that the modern fire world began around this time, 7 million years ago.

Evolution of fire traits

Plants have evolved a number of traits that help them live in a variety of conditions or habitats. It is probable that a trait that evolved in response to some other environmental pressure may be found to be useful in a fiery landscape. For example, having a thick bark may provide a tree with a number of advantages as it protects the outer cambial growth layer of a tree or shrub. There is no doubt, however, that a thick bark layer is particularly advantageous to some conifers, such as pines, which experience regular fire. Such a trait may be thought of as a fire trait. Other traits that plants have developed that have been associated with fire include re-sprouting, serotiny (an ecological adaptation by which seed release is predominantly in response to being heated by fire), and germination in response to heat or smoke. However, it has been suggested that these traits are not simply an adaptation to fire but rather to a fire regime. A fire regime includes characters such as fire frequency and fire intensity, as well as patterns of fuel consumption (Figure 11).

Recent molecular analysis of a range of plant groups including pines and Proteas have suggested that these traits, important to plants living in a fiery landscape, evolved in the high-fire world of the Cretaceous Period, around 100 million years ago. So the traits may be considered adaptive in fire-prone environments and convey a resilience to specific fire regimes (Table 2).

We have already noted that some pines that first evolved in the Cretaceous developed a thick bark that helped them survive frequent fires. However, some taxa also self-prune their lower branches, opening up a gap between litter on the forest floor and the tree

11. The fire regime triangle.

Table 2. Fire traits in plants

Fire-related plant traits	Description	Comment
Thick bark	Fire resistant tissues and self-pruning of branches from tree—adaptation to frequent surface fires, e.g. pines.	Bark thickness strongly influences stem survival. Trees and shrubs with buds above ground are most vulnerable to fire damage.
Post-fire re-sprouting, woody species	Often below-ground re-sprouting from meristomatic tissues followed by post-fire re-sprouting. Epicormous growth such as in *Eucalyptus*.	Eucalypts have buds deeply embedded in the bark and are able to re-sprout even after severe fires. Basally sprouting woody plants can regenerate their whole canopy. Clonal spread is often stimulated after removal of above-ground stems.
Fire-stimulated germination	Often found in fire-prone systems such as chaparral.	Heat shock germination is common in a wide range of flowering plant species especially from Mediterrannean-type climates.
Smoke-stimulated	Smoke-stimulated flowering such as in some Proteas.	Found in a large number of plants including many angiosperms.
Serotiny	Canopy seed storage and fire-stimulated seed release. Often associated with crown fire regimes.	Found in many conifers including *Pinus* but also in some southern hemisphere taxa of both conifers and angiosperms.
Pyrophytic annuals	Regular germination following fire.	
Fire-stimulated flowering	Effective in synchronizing recruitment to the post-burn environment.	Found in many monocot angiosperms including grasses and orchids.

canopy by removing ladder fuel. Other pines, especially those living in higher altitudes such as jack pine and lodgepole pines, do not possess thick bark but have evolved a cone that opens only after a high-intensity, stand-replacing fire and the seeds are then released.

Another adaptation that some trees have evolved is re-sprouting (Figure 12). Such is the case with eucalypts in Australia, which may experience high-intensity crown fires but have developed a technique of re-sprouting from different parts of the plant.

Some plants, such as Proteas, many of which thrive in the Cape area of South Africa, have developed a sensitivity to smoke whereby they sense a coming fire and produce seeds that are released only after the fire has passed. Dormant seeds in the soil seed bank may also sprout following a fire.

While ascribing a particular plant trait to fire may be problematic there is no doubt that fire has been a major pressure in some

(a) (b)

12. Fire adaptive plant traits. (a) Re-sprouting; (b) serotiny.

ecosystems throughout geological history and adaptations to one environmental factor may also have been useful and selected for because of fire. A good example is clonal growth. This is a habit that first evolved in the earliest land plants but is a particularly useful trait in disturbed environments, even for example in volcanic terrains. Horsetails, a group of plants that became so important through the Carboniferous and after, developed this strategy and exhibit it today, much to the despair of gardeners trying to eradicate them from their garden. Yet clonal growth is useful in fiery regimes. One of the largest clonal plants is the quaking aspen (*Populus tremuloides*; it is in fact the world's largest plant in terms of biomass and also one of the oldest), which often thrives following fires, and is especially associated with pine forest fires.

An increasing number of studies now recognize the importance of fire in the ecology of a number of different vegetation types. We may think of different communities of plants, some of which may be totally destroyed by fire but others of which may not only survive fire but in some cases need fire. Fire cannot and should not always be excluded from the landscape. The integration of fire studies with traditional plant ecological studies has provided significant debate. We now have a better understanding of the role of fire on Earth, and that has proved fundamental to conservation policy. It is now appreciated that many of the grasslands of Africa are ancient and are maintained by fire rather than simply be degraded forest. Excluding fire for some ecosystems may have a devastating consequence for some habitats. Fire, then, may be thought of as an ecosystem process and fire regime may be considered as the complex interaction of primary productivity, seasonality, ignition source, and fuel structure. One element that is key is the fire return interval. It is possible for one type of vegetation to be completely destroyed by a fire where there are rare or absent fires, such as in tropical rainforest. In contrast, some grasslands thrive on fire return intervals of 1–25 years.

A change in fire frequency even in vegetation that is well adapted to fire may have catastrophic results.

Fire and animals

While we often consider the relationship between fire and plants and the increasing acceptance that fire has shaped many plant adaptive traits, the relationship between fire and animals has been explored far less. Juli Pausas of the University of Valencia, Spain, has documented a wide range of interactions of fire and animals, though this type of study is in its infancy (Table 3). He suggests that fire is an important evolutionary driver for animal diversity. This is not an aspect that has been explored in the fossil record.

Pausas's hypothesis is based on the observation that many animals present today in fire-prone landscapes have characteristics that contributed to their adaptation to open environments and also that in some cases animals inhabiting fire-prone ecosystems may show specific fire adaptations (Figure 13).

In 2017 one such example hit the newspaper headlines, when it was shown in Australia that some birds of prey may even have started fires. Intentional fire spreading by several raptors such as the Black Kite, Whistling Kite, and Brown Falcon has been reported in Australian savanna grasslands. The birds have been observed to grasp burning twigs in their talons or beaks, apparently to encourage the spread of a fire so they may have easier access to their prey. Many animals, such as insects, can also thrive following a fire.

Pyrogeography and pyrodiversity

The increasing use of satellite data has transformed our understanding of fire and shows just how widespread and common fire is on Earth. This revolution has led to a broader recognition of the special and temporal patterns of fire, which has

Table 3. Possible benefits to animals of fire and fire-altered habitat

Benefit	Category	Fauna
Fresh grasses, and leaves	Food resource	Herbivores, e.g. large mammalian grazers, insect herbivores, arboreal marsupials
Fire-released seed, and more exposed seeds in the soil	Food resource	Granivores including rodents, seed-removing ants
Animals fleeing or dying	Food resource	Predators, scavengers (e.g. birds, kites, owls, ants)
Weakened and dead trees	Food resource	Bark beetles, cavity-dependent (hollow-nesting) animals like woodpeckers, other birds, lizards, possums
Dead wood	Food resource	Saproxylic insects
Flowers, post-fire blossom	Food resource	Insect pollinators, hummingbirds
Meeting point	Mating cue	Saproxylic insects, smoke flies, mole crickets
Synchronization of the emergence	Mating cue	Insects (some beetles)
Reduced habitat complexity: increased visibility	Habitat alteration	Birds of prey; large herbivores, primates (easier to move and detect their predators)
Reduced habitat complexity: movement through the environment	Habitat alteration	Grouse (gaps for mating); seed-dispersing ants (move further with fire)
Microclimate change	Habitat alteration	Ectotherms—e.g. thermophilous reptiles, insects (warmer post-fire environment)

Fire

Reduction of parasites	Biotic interactions	Vertebrates
Reduction of predators	Biotic interactions	Insects (e.g. reduction of insectivore vertebrates)

Modified from Table 2 by permission from Springer, Evolutionary Ecology, 'Towards an understanding of the evolutionary role of fire in animals', Pausas, J. G. and Parr, C. L, COPYRIGHT 2018, 32:113–125 https://doi.org/10.1007/s10682-018-9927-6 with permission of author.

led to the new science of pyrogeography that has been championed by David Bowman of the University of Tasmania. It has been demonstrated that there is a significant relationship between net primary productivity (which is controlled by climate) and area burnt. Clearly in those areas that are very dry, such as deserts, the area burned is constrained by the available fuel, which is low because of low primary productivity.

But in areas of high primary productivity such as rainforests the burnt area is also controlled by climate, with rain excluding fire. In areas which have intermediate levels of primary productivity and a high frequency of dry periods, sustainable fire may be more widespread (Figure 14). Such environments include tropical savannas, which are the most flammable widespread environments on Earth. These systems have both abundant fuels and suitable weather conditions to sustain fire. Savannas have hot wet seasons that promote growth and hence fuel, and this is followed by a dry season that helps dry this abundant fuel, and weather that promotes fire (Box 2).

It has been claimed that there may be alternative stable states of vegetation that exist within a region depending on whether fire is present or absent.

An understanding of pyrogeography is proving vital to our approach to conservation as the distinction between 'natural' and 'human-started' fire becomes ever more blurred. As we shall see,

13. Some fire adaptive traits in animals. (a) An emu that blends into a burned grassland. (b) An owl hunting in advance of a fire front.

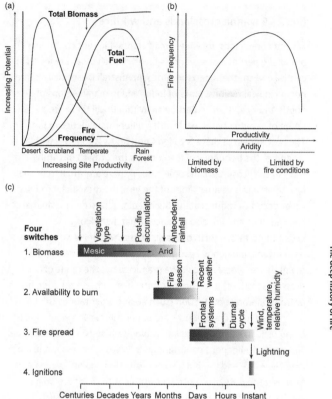

14. Plant productivity, fire, and switches. (a) Increasing fire with plant productivity; (b) Productivity and fire frequency; (c) Fire switches and time.

humans are altering vegetation, and that has a dramatic impact upon fire regimes. A key element in our understanding of fire is the boundaries between flammable (pyrophylic) and less flammable (pyrophobic) vegetation.

To the recent concept of pyrogeography we can add the emerging idea of pyrodiversity, again championed by David Bowman, and

Box 2. Savanna grasslands and wildfire

Savanna grasslands are widespread across different continents but all are part of a diverse grass-fire cycle. Savannas are found in central and southern parts of Africa, in Australia, especially in the north tropical regions, parts of India, and in many regions of South America, Brazil in particular. Within Brazil there are a range of savanna types, each with a different incidence of fire. The best known of these is called Cerrado. As is the case with all savannas, the frequency of fire controls the growth of shrubs and trees in the ecosystem. In some cases if there are small trees that lose their leaves at drier times of the year the grass-rich ground cover becomes more evident. Such regions have diverse animal communities and fire also produces a patchwork of varied ecosystems. In the north of Brazil, on the borders of Venezuela and Guyana, indigenous peoples have successfully used fire to maintain their local agricultural practices, but the range of pressures from other competing activities such as from logging or from rainforest protection (such areas border rainforest) means that local fire practices are under threat even where fire is being used in a controllable and ecologically sensitive manner. Fire use by indigenous communities involves a process not just of land clearance and use but also of protection against even larger fires. Is it the duty of a central government to make fire policy or can local communities with their fire knowledge make such decisions?

Cerrado, which originally covered around 25 per cent of Brazilian territory, is the largest savanna area in South America. Unlike African savanna it is characterized by a layer of grasses but also contains small palms, trees, and shrubs. Like many other fire-prone regions, it has a wet season promoting high biomass (fuel) production and a long dry season. Also in common with other such areas the grasses are highly resistant to fire and rapidly re-sprout after fire. The fires that occur here, whether natural or human set, are surface fires.

A major problem in such regions has become the significant replacement of the native vegetation by a combination of introduced exotic grasses and mechanized agricultural practices that have altered the natural fire regime to more frequent fires, which have a significant impact on the ecosystem.

defined as the coupling of biodiversity and fire regimes in food webs. The importance of feedbacks relating to fire regimes, biodiversity, and ecological processes has been emphasized in this approach. In this context humans can both alter and shape pyrodiversity through the manipulation of fire, in terms not only of frequency of burn and severity of burn, but also of the timing and extent of burning. In addition, fire systems can be altered by the introduction of exotic or invasive species. Clearly any discussion of conservation of biodiversity needs to consider pyrodiversity and an appreciation of the role that fire has played and still plays on Earth.

Chapter 3
Fire and humankind

Wildfire has played a significant role on Earth for over
400 million years not only in shaping aspects of the Earth
System but also on the evolution of its vegetation and
environments. Humans, then, evolved in a fiery landscape and
a key feature of our long interaction with fire has been our
changing relationship with it.

Most now believe that hominins evolved in Africa and indeed that
is where modern humans (*Homo sapiens*) also evolved. It is
evident that these primates would have experienced fire, and fire
would have been particularly frequent in the savannas that began
to appear as the climate dried.

The discovery of fire

The discovery and use of fire has been described by John Gowlett
of Liverpool University as a long and convoluted process. It is
likely that our ancestors' first encounter with fire could be
categorized as opportunistic. A bipedal gait would have allowed
early humans to see smoke from a fire a long way off, and such
naturally occurring fires would have presented several benefits.
Some animals may be driven away by the fire, providing hunting
opportunities. If an animal was killed by the fire then it would
be readily available food and it may have also been cooked.

This is where the idea of cooking meat may have come from. And following the fire, rains might produce a flush of plant growth that would attract grazing animals that could be hunted.

It is difficult to know when exactly humans first used fire. The occurrence of burnt soils within what have been claimed to be hearths comes from around 1.5 million years ago, from the Wonderwerk Cave in southern Africa. It may be that for a considerable time after that humans were able to conserve fire that was encountered opportunistically, for example by adding dung, which is slow burning.

Fire is likely to have been important to early humans for a variety of reasons, and we can consider five major benefits. First, fire would have afforded some protection against large predators. Fire also gives off smoke that may have been useful in warding off insects. In higher, cooler latitudes an obvious use was for warmth, even more than light.

Richard Wrangham has argued, though, that the most significant aspect of fire was its use for cooking. It is not clear, however, when this first took place. There is increasing evidence of the use of fire from 400,000 to 300,000 years ago. Some of this comes from caves in Israel. It has been suggested that cooking may not have been a regular activity until the Late Paleolithic, that is 50,000 to as late as 10,000 years ago. Clearly cooking can be considered a milestone for the use of fire, making meat much more digestible, and neutralizing toxins and pathogens. Cooking may have changed the way food was collected and used. It is not until relatively recently, perhaps less than 10,000 years ago, that cereals were cooked and this had a big impact for human communities. The domestication of crops probably originated in western Asia, although an exact date is difficult to determine. It did not immediately spread across the Middle East and into Europe where we don't have evidence of charred grains until around 7000 BCE.

Another critical development in the use of fire was probably occurring in the meantime: going beyond conserving fire, and learning to start it. The use of flints to produce sparks to start a fire is not found until around 40,000 years ago.

The development of hearths to conserve fire, initially to ward off predators and later for cooking, would have had other important effects. The light would have allowed stretching of the 'working day' and would have been an area around which tool making may have taken place. The heat from the fire may have aided a range of pyro-technologies such as the hardening of spear tips, production of mastic, etc.

Another consequence of activity around a fire would have been to create a social focus. It has been suggested by John Gowlett that this helped group interactions, the development of ritual, and indeed language, and may have played a role in brain enlargement and development.

The utilization of fire in the landscape

Discovering when humans first used fire to purposely change their environment is an almost intractable problem. As we noted, humans were probably able to manipulate fire by discovering and conserving it long before they were able to start fire. Starting a fire involves quite a different process from controlling fire. In the fossil record we may be able to discover from the occurrence of charcoal that there was a fire but it is not possible to determine how and why a fire was started. Given that this may be difficult even in modern times it is extremely difficult to know how this may be done for fires in the past. There have been two approaches to this problem, both involving fire records based upon charcoal analysis from drill core sequences. The first approach is to examine long-term records of fire activity to discover whether there are changes that could not be explained simply by climate changes. A serious difficulty with such an approach is that humans may introduce fire to vegetation that already experiences wildfire. Humans may change when fire occurs in a year and this would not be picked up in the charcoal record.

A second, potentially more successful approach, though it is not without its critics, has been used for example in considering changes in fire activity in Australia that might be associated with the influx of humans. We know that Aboriginal peoples developed a fire-stick culture, and a rise in fire activity some 50,000–60,000 years ago has been ascribed to this development.

Sally Archibald and colleagues from South Africa have claimed that human populations were already shaping landscapes by fire by the African Middle Stone Age, around 100,000 years ago. However, one way to determine when the widespread use of fire to alter the landscape arose is to look at the change from hunter-gatherer populations to agricultural and pastoral populations from around 10,000 years ago. The advent of farming also provided for a rise in population numbers but this is relatively recent in terms of Earth history.

We may think of landscape modification by fire as simply resulting in forest clearing, for example in slash and burn agriculture. However, the use of fire may be more subtle; in some cases fire may even be used to increase the crop of nuts, such as acorns and chestnuts, from trees.

The long-term impacts of fire use and management are also seen in North America where, especially in the south-west, native American populations often undertook regular surface burns, often to drive out game animals. The suppression of such activity has had a significant consequence for fire occurrence and severity in the modern day, as we shall see.

The use of fire to modify the landscape remains a divisive issue and attitudes to such use vary with geography and culture.

While in many advanced Western cultures there has been a significant move away from the use of fire on the landscape, in other parts of the world there are two quite different approaches or views that have very different consequences.

The use of slash and burn agriculture in biodiversity sensitive areas, such as in the Amazon rainforest, has brought widespread condemnation of this practice on a vegetation that does not normally experience fire, which degrades an important global resource (Figure 15).

Yet in other areas of the Amazon Basin, for example, such as in parts of Venezuela, Brazil, and Guyana, some indigenous communities have developed an understanding of fire and fire management that does not promote habitat loss. In many cases, however, a 'one size fits all' policy of fire suppression may not be the answer.

15. Slash and burning of rainforest in Brazil.

However, the pressure to extend land use in some areas, such as Brazil, is a cause for concern. This takes two forms. The first is forest clearance and its replacement with animal grazing pasture, or in some cases to develop arable land. Pressure may also arise for the development of such areas due to population growth. The second is for logging, with large trees felled and the slash burned.

The consequence of this land use change is nowhere better demonstrated than in Indonesia. Again, there are two competing needs. The first relates to the logging of tropical forests that do not normally experience fire. In the case of many of the Indonesian areas such as in Kalimantan (Borneo) and Sumatra logging has given rise to a number of unintended consequences. This is because many of the trees are growing on wet peats. After logging the peat may dry out and the resultant dry peat provides an ideal fuel for a fire. The problem is exacerbated by the El Niño effect, by which every seven years or so such areas experience severe dryness. In 1982/3 there was a fire across an area of Kalimantan that was the size of Belgium, Holland, and Luxembourg combined.

In addition, other peat-land areas have been drained and logged for the development of oil palm plantations. This too has led to the occurrence of major fires that have global consequences such as the production of large quantities of carbon dioxide, the greenhouse gas, as well as smoke that can travel thousands of kilometres, as demonstrated by Susan Page of the University of Leicester. And as Fay Johnson of the University of Tasmania has shown, such smoke can have serious human health implications.

Capturing and controlling fire

The capturing and utilization of fire can be seen as a defining human trait. But capturing, utilizing, and controlling fire represent quite different activities. As we have seen, the opportunistic capturing of fire by early humans did not necessarily lead to extensive use. The ability to maintain a fire helped its development for human use, but it was not until the ability to start fires was achieved that human uses could fully develop. Using fire is not the same as controlling fire: controlling fire requires not only knowledge of kindling and sustaining it but also of how to extinguish it. In conditions such as an open hearth the combustion of fuel that is not replaced will lead to the eventual extinguishing of the fire. Dousing a fire with water may have also been learned early in the development of fire technologies.

The next steps in fire utilization came from an increasing understanding of how different fuels may burn, both in terms of types of plants, and their size and condition (wet/dry, etc.). Within a landscape it is necessary to develop an understanding not only of how to burn but also when to burn. Partial control of fire may be more possible where it is started outside a normal fire season when, for example, the plants are not fully dry. We can see this happening in central Africa today when fires are set before the main dry season begins. This was also the approach of native American populations, who started small surface fires in forested areas that had the additional benefit of preventing larger crown fires during drier periods by reducing surface fuel loads. Stopping such practices has had significant consequences for the size and intensity of forest fires in the south-west of the USA, as shown by Tom Swetnam, Chris Roos, and their colleagues from the University of Arizona and Southern Methodist University in Dallas, Texas, respectively.

Patch burning as part of the fire-stick culture of Australian Aboriginal populations may also have had a possibly unintended beneficial consequence. It produces a mosaic of burned and unburned areas on the landscape that have been shown to limit the size and intensity of some natural lightning-ignited fires.

While we may easily start a fire, controlling and extinguishing it is far from easy, as we have seen in many parts of the world over the past few years. We will consider these aspects more fully later in the book.

Fire use in agriculture

A major change in the use of fire arose when it began to be more fully controlled and used in a range of agricultural practices. The first way in which fire had an impact on the land was not from the fire itself but from the manipulation of the fuel. This may have involved everything from the draining of an area to cutting down woods or driving animals to trample the fuel. The timing of the burn and the size of burn allowed have also been important factors. Natural fire systems have given way to human-managed fire systems. Other aspects too can be manipulated. The vegetation itself can be changed, with new species introduced with a range of different flammabilities. This has occurred not least with the widespread introduction of cereal crops. We can consider examples of agricultural or pastoral fire practices from around the world, to demonstrate their variety.

Our first example has been called a 'fire–fallow cycle'. In such a system fire plays a major role. The term that agronomists use for fuel is fallow. If a field is left for several years then it may produce a range of weeds, shrubs, and even small trees and plant invasives. In some systems, especially in Europe, these fields were seen essentially as waste and they were often just burned. However, some saw this as a neglected asset and the idea was introduced that saw such burning as a productive opportunity. The burning had a significant ecological impact on the field, not only reducing pests and weeds but also returning nutrients to the soil and indeed modifying soil structure.

One of these forms of agriculture, known as Swidden, is still widely practised world-wide. This is a technique of rotational farming in which land is cleared for cultivation (normally by fire) and then left to regenerate for a few years. A variation has often been termed slash and burn. In this case an area of vegetation is identified for growing crops, and the trees and shrubs are felled, then allowed to dry and subsequently burned. In such a system, in the first year following the burn, the crops thrive; but by the second year after the burn the crops may struggle to yield, and improvement in productivity may require a further burn. The third year sees a return of the native flora in a normal sequence of post-fire recovery. Whereas such a system may prove useful in some places, in others it requires abandonment and moving on and it is this practice that has led to an overall resistance by many conservationists, especially in areas such as the Amazon Basin where fires do not occur naturally and the primary forests are a major carbon sink and oxygen producer. But this is not the situation in every case as fields can be revisited and the burning cycle repeated, and the fire may become much easier to control. Despite this practice being considered 'bad' in many cases, it can, over centuries, create a mosaic landscape which has a range of patches in different states of cultivation and fallow that in turn may encourage biodiversity and indeed provide a range of animal and rare plant habitats. It leads to the new concept that 'pyrodiversity begets biodiversity'.

The 'slash and burn' technique can be refined into a more regular field rotation practice. In such cases the plots remain much more fixed in their position and size. Here, following a burn, there is a distinct sequence of crops that may be planted to help in the development of soil fertility, such as cereals followed by root or tuber crops. This may be a cycle ranging from two to even eight years, after which the field is again burned. Field rotation will mean that in a given area some fields will be under different crops but others readied for burning. This then is a 'natural cycle' that does not involve continuous farming using intensive treatments such as continual tilling and fertilizing.

While such practices of field burning will break up the landscape, reducing the likely spread of natural lightning-started fire, on the other hand, they also result in significant smoke pollution, which has now been recognized as a health hazard.

What may be surprising is the range of areas where fire-fallow farming has been followed. This includes tropical areas such as in Thailand and central Africa, and even in more temperate regions such as northern Europe. Each of these regions provides different challenges for both farmers and conservationists alike.

Our second example we can call fire-forage pastoralism. This is a system that manages fire and flora for the benefit of animals. The idea is that the burning allows the renewal of the landscape that favours vegetation used by either grazers or browsers. In particular it has an impact on grasses. In some areas where, for example, C4 grasses grow, these may not be palatable when old but if burned the new shoots provide a good edible crop. But regular fire may also be needed to prevent an area returning to woody scrubland or forest. And we can consider fire as a method of rejuvenation just as we saw with the fire-fallow farming.

In the pastoral setting, however, we see an additional tension. William Bond and Jon Keeley considered fire as a 'global herbivore'. This implied that fire and herbivores are competing for the same food resource and in some respects they have the same impacts—combustion, albeit one slowly, through eating it, and the other quickly, through burning it. Both have an impact on the vegetation itself and both produce waste products that include solid residues and greenhouse gases (methane and carbon dioxide).

In a natural environment fire will occur by chance, often depending on weather, ignitions, and essentially timing being unpredictable. In a fire-forage pastoralist system there is an element of order introduced into the system allowing regularity to both the burning and herding. But this practice does not always sit well with those wishing to undertake more settled arable systems and this clash adds to the complexity of community and political attitudes to fire.

There are many approaches to the use of fire in these pastoralism examples. One of these has been termed 'transhumance'. This is where, in mountainous regions, forage is available at different seasons in the year and is particularly common in the Mediterranean Basin. Animals are moved into the mountains for summer grazing and back to lowland areas in the winter. In many cases the mountain pastures and also the migration routes are burned in advance of the movement of animals allowing for a flush of new edible growth. Also, the areas are burned at the end of the season. Even the winter pasture may be burned to help eliminate shrubs and small trees that may have grown up in the summer months while the animals were away.

In other parts of Europe, in more temperate settings, attempts have been made to integrate animals with crop agriculture. Animals spend their time outside in the summer but inside during the winter. In such cases fire can be used in rotation in different plots to 'freshen' the fields, and manure from the animals kept indoors can be put on the fields to enhance growth.

Such mixed fire/agricultural practices are found around the world as far afield as Sweden in the northern temperate regions, to Madagascar in the more tropical region of the southern hemisphere.

Industrial use of fire

It is believed that two glues in use by 100,000–50,000 years ago required fire. Twine needed for working a fire drill appears to have been in use 120,000 years ago. The first use of what we may call industrial fire is likely to have been in the making of pottery. This craft is thought to have developed in China; exactly when is debated but it could have been as early as 20,000 years ago.

Pottery vessels could be used not only for storage but also for cooking. It is only about 5,000 years ago that we see the true industrialization of fire with the development of metal working, which first involved copper and bronze and then iron. Metal smelting is a complex technology that involves developing a persistent hearth with high temperatures rather different from a usual open hearth fire. Additionally we now have information from the ice core records in Greenland that suggest the smelting of lead and silver was being undertaken 5,000 years ago. The main impact of these developing technologies came with the expansion of the Roman Empire.

We have seen how open fires can be used as a source of heat. The development of the Roman hypocaust system (a heating system used in Roman homes and baths designed, rather like a modern central heating system, to circulate hot air under the floors and walls) allowed for an increase in the distribution and use of heat. The heat could be used not only for under floor hot air heating but also for heating hot water for both private and public baths. The significance of the development of hypocaust systems lies in both fuel production and use. While we have many examples of Roman hypocaust systems we have little data on how they were fired or the temperatures used. Research on the hypocaust systems of Pompeii has shown the large areas of forest that would have been needed to provide the fuel. It has also been suggested that a more efficient way of moving the fuel and reducing smoke in the urban environment would have been to produce charcoal out of the city and transport it in for use. We know that charcoal was in use in Egypt several thousand years before. Charcoal use would not only reduce smoke emissions but provide higher and more even temperatures in the ovens. We have, however, little data of the fuels that were used for hypocausts in general, as archaeological investigations usually clean out the systems and there has always been difficulty when charcoal is discovered to distinguish which of these charcoal residues were from the burning of wood, as opposed to the partial combustion of charcoal that was used as a fuel.

A technique involving measuring the reflectance of the charcoal, developed by myself and my research students for the study of charcoal, allows us to make such a distinction. This has revolutionized these archaeological studies. For example, in a collaborative project with the charity English Heritage, we were able not only to demonstrate from testing charcoal found in a British Roman hypocaust that wood was used in the hypocaust rather than charcoal, but also to identify the type of wood being used to fire the hypocaust system.

The need to develop higher temperatures in furnaces for working a range of different metals, and also for glass making, had two impacts. The first was in the development of charcoal production. The combustion of charcoal provides much higher temperatures than can be achieved with the combustion of wood alone. This is particularly important for the production of glass. However, a second and major change in industrial processes was driven by the use of other materials such as coal for combustion. The origin of the use of coal is uncertain but by Roman times coal was in widespread use for industrial processes including iron smelting. In the British Isles, for example, there is widespread evidence of coal production, transport, and use. The use of coal revolutionized iron production and led, by the 19th century, to steel production.

Using coal rather than wood as a heat source also had a number of other consequences. Coal could be more efficiently used for steam production for a range of technologies including engines both for use in factories and subsequently for use in transport and in the home. We shall look at this change from a rural, agricultural use of fire to urban, industrial applications in Chapter 4.

Fire as a weapon

The increasing use of fire on the landscape and indeed of fire in a domestic setting often led to a controlled fire becoming an uncontrolled fire. Fire may spread from a wildland area into an area with buildings, which in many cases were flammable. In addition, the fact that an open fire was often found within a building led to the distinct possibility of its spread, leading to the accidental destruction of property and loss of life. As fire can be so destructive it also has the potential to be a weapon of war.

Over 50,000 years ago, the hafting of a Levallois point, a specialized flint core which has been knapped to remove thin flint shards and worked to a point, involved the use of fire. Fire was also used to harden the points of spears. The ability to use fire as a weapon may have developed from the throwing of flaming spears or arrows. Extinguishing large fires especially in flammable buildings can be very problematic and their ability to spread quickly led to the extensive use of fire as a destructive force. Not only can fire be used to destroy as an attacking weapon, but retreating armies have used it in a 'scorched earth' policy, to deny advancing armies food or shelter.

Technologies needed to be developed to move fire safely away from an attacking force while being used to destroy a defending force. Finding materials that would burn for lengthy periods is important in this regard. Bitumen was readily available in the classical world especially in the Middle East and may have coated arrows or more likely have been impregnated into cloth that could be wrapped around projectiles and fired at an enemy. It has been suggested that such weapons were used by the Assyrians as early as the 9th century BCE and in the 5th century the Greeks were recorded as using tubes which blew large flames. We know that the Roman emperor Septimus Severus constructed a bath that was fuelled with bitumen or petroleum, so such materials must have been available for warfare. Around 670 CE, Greeks besieged in Constantinople used some kind of flammable liquid for what was termed 'Greek Fire'. Unfortunately we still do not know what exactly it was, but some believe that it was made with a mixture of naphtha and quicklime (other ingredients have been suggested), and it could continue to burn floating on water. We also know that charcoal was known in the classical world and flammable liquids are one possible by-product of the charcoal-making process so it may be that this was the source of such a material.

Fire, then, is a method of destroying an opponent, particularly by burning down living quarters as well as administrative and cultural buildings such as temples, as was done by the Greeks when destroying Troy. We tend to imagine that the sacking of a city even several millennia ago resulted in destruction by fire. The problem, however, is to distinguish in the archaeological record between fires that were accidental and became out of control and spread, and those that were started deliberately by an attacking enemy.

Fire could be an effective weapon on land, but it could also be used at sea, where burning projectiles could be fired from one boat to another, and the combination of construction materials and pitch used for waterproofing made early ships extremely vulnerable.

The major change in the use of fire in weaponry came with the discovery of gunpowder in China, documented by Joseph Needham of the University of Cambridge in his monumental account of science and civilization in China. Gunpowder is a mixture of charcoal, sulphur, and potassium nitrate, which was known as saltpetre. In this case the charcoal and sulphur are the fuels and the potassium nitrate acts as an oxidizing agent. Not only does the gunpowder generate heat but also large quantities of gas that can be used as a propellant. Although gunpowder was invented in China in the 7th century CE it was not until the 13th century that it spread to Europe, where its use in weapons was more fully developed. While we think of gunpowder as a material to be used in warfare, it was originally developed for medicinal purposes.

It was not until the late 16th century onwards that the use of gunpowder technologies became more widespread. In the later part of the 19th century, high explosives were developed, such as dynamite and mixtures that included ammonium nitrate and fuel oil.

Through the 20th century a larger number of explosive materials and technologies were devised together with ways in which their destructive materials could be delivered. Initially simple guns or rifles were used and cannons, where the explosive material was essentially used as a propellant. However, developments in the 20th century allowed for the increasing use of exploding shells and also incendiary devices through which fire could be spread much more effectively.

The development of flying machines, particularly aircraft that used combustion technology for flight, became an important way in which to distribute explosives via bombs or incendiary devices to set fire to buildings, which not only had a destructive effect but also tied up personnel to attempt to extinguish the fire as well as having a psychological impact. This was seen in the bombing of London and Coventry among other cities in the Second World War and even more dramatically in the incendiary bombing raids by the Allies on the city of Dresden in Germany in 1945.

The widespread use of fire-producing chemicals came with the extensive use of napalm, developed in 1942 but used during the Vietnam War by the United States in the 1960s and 1970s. Napalm is a highly flammable sticky jelly that is used in incendiary bombs and, in addition, flame-throwers, increasingly used in the 20th century both by human operatives or via mechanized transport (Figure 16). Napalm consists of a mixture of thickening and gelling agents, including co-precipitated aluminium salts of naphthenic and palmitic acids and volatile petrochemicals such as petrol (gasoline) or diesel fuel. In fact military flame-throwers have also been used in forest fire management, and the gelled fuel used in the drip torch or

16. **Fire and warfare: modern use of flame-throwers.**

heli-torch and ground-based devices such as the terra-torch is essentially napalm.

Fire has also been used to help destroy a country's economy, as well as its people and infrastructure. Perhaps the best example of this was seen during the first Gulf War in 1991, in which the Iraqi army of Saddam Hussein set fire to the oil wells as part of a scorched earth policy while retreating from Kuwait. This had political, economic, and publicity impacts. Iraq claimed the oilfields, so by destroying the wells, it sent both a political message and an economic one. In addition, setting fire to the oilfields not only prevented Kuwait from using the oil but took up valuable resources to put out the fires. The smoke plumes from the fires were very visible, showing that the army could create havoc by setting light to perhaps over 700 wells. Extinguishing the fires took more than seven months.

Chapter 4
Containing and suppressing fire

We are all aware that around the world there are different climates and different types of vegetation and that they are somehow linked. We use terms such as tropical rainforest, temperate forests, boreal forests, savanna grasslands, but each of these biomes, as they are called, has a community of plants that have common characteristics suited to their environment and that evolved in response to a shared physical climate, and the plant communities have different susceptibilities to fire. In some regions, such as tropical rainforest, the plants do not regularly experience fire and are very sensitive to it. In contrast, some regions with what we call a Mediterranean climate (wet winters and long dry summers), found not only in the Mediterranean area but also in the western United States of America, parts of South America such as Chile, and parts of Australia and southern Africa, have vegetation that regularly experiences fire and many of the plants have adapted to a fiery regime. But the change from one biome to another can occur over a distance of only a few hundred metres, and one region may contain several biomes. As each biome may have different flammabilities and indeed may or may not need fire, formulating fire suppression policies can be quite complex and may be little understood by the population at large. How is it possible for a fire to be vigorously suppressed in one area yet be allowed to burn in another?

Landscape fires

As the cost of suppressing landscape fire has spiralled over the past decade there are a number of issues we need to consider in regard to whether or not all wildfires are bad and should be suppressed. For example the fire suppression costs in the United States of America have been steadily increasing so that they now represent 50 per cent of the annual budget of the federal government's Forest Service. The Thomas Fire north of Los Angeles in California that threatened and destroyed many homes, including those of several Hollywood stars, in 2017 alone cost 177 million dollars to suppress.

To make decisions about fire suppression, we need a greater understanding of how fires start and spread; we can then make predictions of both these aspects as well as developing appropriate strategies. Fire behaviour is intimately tied up with the nature of a biome, but human influences may also have changed our approach to fire intervention or suppression.

In the early 1950s Jack Barrows, who worked on the northern Rocky Mountain forests, wrote on the behaviour of fire—the manner in which a fuel ignites, flame develops, and fire spreads and exhibits other related phenomena. He outlined five steps involved in the practice of predicting or forecasting wildland fire behaviour: basic knowledge; forest knowledge; aids and guides; estimation of the situation; and decision.

As part of the first step, basic knowledge, we need to consider the fundamentals of combustion. We have already seen, using the fire triangle, that three elements are involved in fire: a fuel to burn, a source of heat, and availability of oxygen. We can develop this concept by considering combustion as occurring in four phases. The first may be called the pre-heating or pre-ignition combustion phase, in which an unburned fuel may be altered by the heat from an advancing fire front. The heat may drive off moisture from the fuel and raise the temperature to ignition point. During this phase, the cellulose in the plant tissues begins to break down, releasing flammable gases.

The second phase is the flaming combustion phase. Here the flammable gases that have been released mix with oxygen in the atmosphere and an oxidation reaction takes place that is essentially an exothermic chemical chain reaction producing heat and light.

The third phase is the smouldering combustion phase, and it is more important than often recognized. This is where the combustion reaction is sustained by a low heat, and the oxygen directly attacks the solid fuel. There are no flames. During this phase, the combustible gases and other volatile vapours continue to escape into the atmosphere, often as visible smoke.

The final stage is the glowing combustion phase. By this time most gases have been driven off, but the carbon-rich fuel continues to oxidize, producing a significant amount of heat with only embers visible and no flame. We are familiar with this form of combustion in charcoal on a barbecue.

Although it is clear from first principles how a fire may start, we need to consider how a fire may spread by not only thinking about the transfer of heat, but also appreciating the different ways in which it may spread (Figure 17). This is what has been termed forest knowledge. When discussing fire spread, different parts of the fire are usually described as the head or front, the flanks, and the rear or back of the fire. Each part of the fire may have different characteristics. The term fire perimeter is used to describe the boundary or outer edge of the fire or burned area, though unburned 'islands' may exist within the fire perimeter. In addition, wind-blown 'spot fires' may occur in advance of the main fire

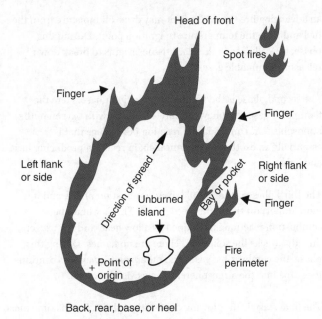

17. Geometry of wildfire spread.

front, even at distances of several kilometres, depending on the nature of the vegetation and the prevailing wind.

In all our discussions of fire the three key elements must always be present for it to exist: there must be sufficient fuel to burn, and appropriate size and arrangement for it to burn; the fuel must be sufficiently dry to support a spreading combustion reaction; and there must be an agent of ignition. Often this last factor is the only one considered when a major wildfire is reported in the news.

We know from experience that having a continuous layer of surface fuels is important for the spread of a fire. This could be dry conifer needles, but also other leaves, twigs, and particularly cured grasses. We have already seen how dried grasses can be used for kindling, so this element should not be a surprise, yet it is often forgotten. And the fuel needs to be dry. The moisture content is key and field experiments have shown that fuels need to have a moisture content less than 30 per cent (though in fast-spreading crown fires even live vegetation with moisture levels of 100 per cent can burn).

There are two main sources of ignition for a fuel. The first and most obvious is cloud-to-ground lightning strikes. Millions of these occur across the globe on a daily basis but obviously not all result in a fire. The second, and in some areas the most significant, source is human activity. Fires may be started accidentally or through carelessness, from a discarded cigarette or abandoned campfire, or simply from a spark or electrical fault in a power line, which is what started the Camp Fire in California that destroyed the town of Paradise. Abandoned disposable barbecues form another common ignition source in some places. They may also be the result of arson. For example, in south Wales the majority of grass fires are the result of arson, often by youth and particularly during holiday periods. Significant progress has been made in the forensic identification of the cause of a fire over the past fifty years.

It is clear, however, that if a fire is to start and spread, two fundamental criteria need to be fulfilled. First, irrespective of how a fire is started, there must be enough heat transferred to the volatile vegetation for it to dry out and reach the ignition temperature by the time the flame front arrives (Figure 18). If not, then, for example, there may be a lightning strike to a tree but the fire may not spread. Secondly, there needs to be sufficient fuel so that it is available for consumption to produce a moving flame front. We know that heat can be transferred in the system through convection, radiation, and conduction. These methods of heat transfer are also important when considering building fires.

18. **Speed of wildfire spread. The time from top left to bottom right photo is 105 seconds.**

Fire, weather, and climate

If one of the fundamentals of a wildfire is the fuel that is to be burned then the condition of that fuel is of utmost importance. The lower the moisture content of the fuel, whether living or more particularly dead, the more likely that it will reach the combustion point for a fire to start and spread. Both climate and weather have a part to play in producing conditions for wildfires to spread. In terms of climate, those with higher temperatures will allow for the drying of fuel more easily. However, this is not as important as the length of a dry season; otherwise we would not expect there to be so many fires in northern areas such as Alaska or northern Europe. Despite long periods of drought, not all areas will necessarily burn, especially if there is insufficient fuel.

Even if the climate allows frequent fires, weather plays a vital role. Several elements of weather are relevant here. Perhaps one of the most important, but least obvious, is that of atmospheric moisture (often termed relative humidity), as although there may not be precipitation, a moderate or high atmospheric moisture content will reduce the speed of drying of a fuel. The second factor is temperature: as it increases, the speed of loss of fuel moisture also increases. Cloudiness can play a part here, as it tends to buffer air temperatures and slow down the drying of fuel. A third and important element is wind speed and even wind direction. Wind plays a significant role in both speeding the drying of fuel and in the spread of a fire once it gets started. The wind may be part of a general atmospheric instability that may cause turbulence and thunderstorm activity that may promote lightning strikes, which act as the ignition source. Many of the Yellowstone wildfires of 1988 were started by lightning and the Santa Anna winds in California help the rapid spread of the wildfires in that region.

Fire will not start even with all the right weather conditions and ignition sources if there is not sufficient fuel of the right kind to burn, but the nature of the fuel can also be altered by a number of factors. Multiple years of drought may influence the condition of both a surface fuel and the living vegetation. In addition, insect attack such as by bark beetles may kill a number of trees, adding large areas of dry fuels that may also be affected by a range of pathogens, which may themselves have been spread by insects. One perhaps surprising factor is the previous occurrence of fire, which may have killed many trees, allowing dry fuel to accumulate. This is why it is a mistake to believe that if there has been one fire another will not occur. Many large fires have burned over relatively recently burned areas and this needs to be borne in mind when thinking of rebuilding after a fire.

All these aspects give rise to the idea of fire weather and fire forecasting. They also affect the ability to undertake prescribed burns.

Fuels

We tend to think of all fuels being the same, but different types of fuel exhibit different flammability. We all know this, usually from experience in trying to start a campfire. Setting light to a large log is difficult, if not impossible. If we were a Scout or a Guide we would have learned to collect kindling that is composed of dry leaves, twigs, or grasses that more easily leads to ignition and to only then add larger pieces of dry wood to a fire to allow it to develop. The nature of the fuel is not only relevant for the initial starting of a fire but may affect how large a fire gets and indeed how hot it becomes, and this may have an impact on the soil and on the possible recovery of the vegetation after the fire goes out. This is where the problem of invasive or exotic grasses comes in. Grasses can grow and spread very rapidly and that is why many have been planted, especially as animal feed. But such grasses when dry may produce a significant fuel load that could alter the nature of a fire, allowing it to spread and develop more quickly, and in some cases burn hotter.

We can think of fuels that may exist in four different areas. The most important are the surface fuels as it is these that are most likely to lead to ignition and initial spread. They can consist of dead plant litter that includes leaves and small twigs, grasses and herbaceous flowering plants, coupled with tree seedlings and stumps, shrubs, and downed logs. In areas that have been logged, surface fuel might include slash piles. Ground fuels are those materials that make up an organic-rich soil comprising peat, decayed wood, and tree roots.

The most surprising component of a fuel complex that proves very important is the ladder or bridge fuels that connect the surface fuels with the crown fuels, which form the fourth category (Figure 19). Bridge fuels include climbing plants, tall shrubs, tree branches and flaky bark, snags and hanging mosses. Crown fuels

19. Ladder fuels.

are those materials that make up the tree canopy and principally include leaves or needles and small twigs (live and dead). If the crowns of trees are optimally connected, the fire can spread from crown to crown, and such crown fires spread more quickly than surface fires. In addition, the convection currents produced by this type of fire may suck in air, making the fire more intense. The smoke and convection plume may also transport burning embers downwind of the main advancing fire, resulting in new ignitions in the form of spot fires that effectively expand the area of the main fire.

In addition to the location of the fire, the dimensions and shape of the fuel, its density, total load, and the live-to-dead ratio of fuel elements all play a role in its form and spread. Of course, there may be seasonal variations in the condition of the fuel such as changes in the relative proportion of live and dead material, and moisture content, as well as fuel discontinuities that can serve as barriers to the spread of surface fire (Box 3).

Topography

Yet another fire triangle has been used by fire managers and scientists. This is the fire environment triangle (Figure 5). We have already dealt with two of its sides: fuel and weather. The third component is topography, which can play a major role in the

Box 3. Why do species matter?

We tend to think of specific vegetational or fuel types as all being more or less susceptible to fire. Yet what is also important is what makes up a particular vegetation or fuel complex. Take, for example, the northern boreal forests that are predominantly made up of large areas of coniferous trees. The burning history and susceptibility to fire has been found to vary across this biome. Recent research shows that the differences in the tree species composition in North American and Eurasian boreal forests explain the differences in fires. So, for example, the fires in North American boreal forests typically tend to be high-intensity crown fires. Many of these forests are dominated by a particular suite of conifers, including Black Spruce and Jack Pine. In contrast, the boreal forest fires in Eurasia, where the forests are dominated by Spruce, Scots Pine, and Larch, tend to be low-intensity surface fires, with less severe impacts. This is why it is misguided for politicians to imagine that fire occurrence and methods of control in one region can necessarily be applied to another.

spread of a fire (Figure 20). The elements of topography that influence fire behaviour include not only the elevation of a site but also the steepness of slope, aspect or slope exposure, the shape of the country, and the presence of natural and man-made barriers to fire spread. These are all important factors in the monitoring and prediction of fire spread, in terms of both direction and speed. In part, these elements of topography also influence the nature of the potential fuel and the weather, which may differ across the topography with, for example, rain occurring preferentially on one slope or another and the prevailing wind direction determined by the lie of the land. Local climate effects may determine a variable range in air temperature, relative humidity, wind speed and direction, and precipitation that in turn may have an impact on the resulting vegetation, including species composition and the

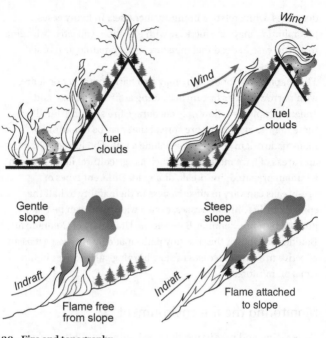

fuel
clouds

Wind

Wind

fuel
clouds

Gentle
slope

Steep
slope

Indraft

Indraft

Flame free
from slope

Flame attached
to slope

20. Fire and topography.

nature of the resulting fuel complex. For example, higher
elevations may experience snow cover and the melting of the snow
in the spring may determine the length of the fire season, which
also depends on how long fuel has to dry and when grass fuels
become readily available to burn.

We can now consider how the behaviour of a fire is affected by
these different topographic factors. As the angle of a slope
increases, the effect is to expose the fuels ahead of the flames
advancing upslope to increasing convective and radiant heating
which in turn leads to increased drying and as a result a higher
rate of spread. Elevation too plays a part. As height above sea level
increases, weather parameters change: air temperature typically

decreases, while relative humidity increases. In many cases precipitation may also increase with elevation, collectively leading to an increase in dead fuel moisture on a climatological basis.

The shape of the topography may also have an impact on a fire, with narrow ravines or canyons acting as fire chimneys, and enhanced uphill air-flow may encourage fire spotting. Finally, there may be barriers to fire spread that may be of natural or man-made origin. These may include a water body, rock-slides, and areas of bare mineral soil (such as agricultural fields) or even a human prepared fire-break, or a road. Different types of fire-breaks can vary in effectiveness in their ability to halt the spread of a fire. In some cases, even a wide break can be readily jumped by a fire. Jennifer Balch at the University of Colorado at Boulder has shown that the tiny paths made by ants can act as an effective fire-break in cases of slow burning surface fires in the Amazon rainforest.

Monitoring the fire environment

Forecasting and predicting the possibility of a wildfire is not an easy job. The third step indicated by Jack Burrows relates to aids and guides. A number of aids and guides have been developed in different countries to help with this. While some of these may be appropriate for one region new aids and guides need to be developed for others. There are so many variables that need to be considered, and different regions of the Earth with different vegetation types require different approaches. So fire warning systems developed for Canada or the USA, for example, may not be appropriate for Australia, Spain, or the UK.

Once a fire has started, many factors come into play in the prediction of its behaviour. Fortunately we now have available several means by which to monitor an active fire not just from the land and air but also from satellites. Together with detailed

Global Positioning Systems (GPS), Light Detection and Ranging (LIDAR), and optical remote sensing, all coupled with great technological advances in computing, we can track the progress of a fire on a near real-time basis in some cases. In addition, we can now add real-time data to the system that predicts how a fire will develop. However, even with this capability it may still not be possible to accurately predict the behaviour of a fire with predictive systems if there are errors in the input data or errors in the modelling of fire behaviour, which could have disastrous consequences for the public and fire fighters.

Wildfire behaviour

Various means have been used to describe a wildfire's behaviour, such as cool versus hot or light versus severe, that may be more qualitative than quantitative in nature. Descriptive terms such as smouldering, creeping or running, torching, and spotting are also commonly used. But regardless of the terminology used, all wildland fires exhibit a number of fundamental characteristics—they spread, consume fuel, and produce heat, light, and smoke.

Wildfires may be propagated in a number of different ways, either as a ground or subsurface fire, a surface fire, or a crown fire (Figure 21). The distinction is important as while a crown or surface fire may be extinguished, the underground, often smouldering, ground fire can continue to burn and erupt into a new surface or crown fire after the main fire appears to have been extinguished. In many forested vegetation types it is possible that all three fire types may occur at the same time but extinguishing each type may require different approaches. How a fire starts from a single ignition source (e.g. by lightning or a campfire) or line source (such as fires started deliberately as part of a prescribed burn) will have an influence on how it spreads from its ignition source. And as we have noted, exactly how a fire spreads can be complicated by changes in fuel type and condition, topographic

(b)

21. Surface to crown fires. (a) Surface fire in savanna; (b) Surface fire transition to crown fire with ladder fuels alight.

features, and, in particular, weather conditions. While extensive efforts may be made in extinguishing large fires they often halt because of a change in weather such as a drop in wind speed and/or temperature, an increase in relative humidity, or the occurrence of rain.

Having taken all of our basic and forest knowledge into consideration and using all the aids and guides at our disposal we need to make an estimation of the dangers a fire poses in order to come to a decision on the best method to suppress it or indeed to let it burn. The distinction between fire types is important, therefore, in decisions of how to tackle a fire and indeed whether or not to fight a fire in the first place. Many surface fires are cool and slow moving. The fire-line intensity is formally defined as the rate at which the consumption of fuel produces heat per unit of time per length of the fire front, and in reality can be estimated by combining the fire spread rate and fuel consumption. Slow-moving surface fires with little surface fuel will not have the same fire-line intensity as surface fires that consume large quantities of fuel. This may be significant as regular surface fires may reduce surface fuel loads and prevent a higher-intensity fire in the future. This is because intense surface fires can initiate crown fires, depending on a range of factors including foliar moisture and canopy base height. Once a crown fire has developed it may separate from the surface fire and a much more dangerous situation can develop, including the increase in fire spotting and overall destruction of vegetation.

In addition, extreme fire behaviour can develop through sucking air into the fire and the sudden extension of a fire (blow up) and even the formation of 'fire tornadoes'.

Rates of fire spread may depend on the nature of the vegetation, fire type, and in particular wind speed. In some cases fire fronts have been known to move as fast as 30 km/hour for short periods of time but usually the 10 per cent rule applies, meaning the fire moves at a rate of about 10 per cent of the wind speed. Wildfires in some forests have been recorded to run for upwards of 65 km in a single direction driven by the wind over a ten-hour period. This knowledge is important for those planning evacuation of habitation in advance of a fire.

Preventing and extinguishing wildfires

Wildfire is a natural phenomenon in many ecosystems. Indeed both the climate and vegetation can combine to make some ecosystems particularly flammable. We should not, therefore, consider all fires 'bad' and requiring to be extinguished. Frequent low-intensity surface fires in some forested ecosystems may reduce surface fuel loads and help prevent the development of more dangerous crown fires. In some cases the intentional setting of surface fires may be useful, and it is a practice long carried out by several indigenous populations across the world. In other cultures prescribed burning may be an option to prevent the occurrence of more dangerous fires. Even so, it is not without risk, as the fire may 'escape' from a managed prescribed surface burn to an out of control wildfire, especially if the weather changes unexpectedly. Nevertheless, it is unfortunate that misunderstanding of prescribed burning has led in some cases to public and governmental rejection of such a practice, as it may play an important role in the land manager's armoury against wildfire (Table 4).

All of our understanding of wildfire behaviour must be brought to bear when we consider how to extinguish a wildfire. (If that is the plan: it may be that a 'let it burn' policy is best as long as population centres and valuable infrastructure are not threatened). Decisions need to be made on how difficult it is to control and put out a fire and what equipment and personnel will be needed.

The choice of methods to be used will involve the prediction of fire behaviour. Models of fire behaviour vary in their complexity and may draw on knowledge derived from chemistry and physics, from field observations, and from laboratory experiments. Some also incorporate atmospheric processes. Depending on the predictions of a model, choices will be made over the use of one of the ground fire fighting methods and also for aerial 'bombing'.

Table 4. Human influence on fire

Fire variable	Natural influences	Human influences	Fire regime parameters
Wind speed	Season	Climate change	Fire spread
	Weather	Land cover change	
	Topography		
	Land cover		
Fuel continuity	Terrain type (slope, rockiness, aspect)	Artificial barriers (roads, fuel breaks)	
		Habitat fragmentation (fields)	
	Rivers and water bodies	Exotic grasses	
	Season	Land management (patch burning, fuel treatments)	
	Vegetation (type, age, phenology)		
		Fire suppression	
Fuel loads	Tree, shrub and grass cover	Grazing	Fire intensity and severity
	Natural disturbances (e.g. insect or frost damage, windthrow)	Timber harvests	
		Exotic species establishment	
	Herbivory	Fire suppression	
	Soil fertility	Fuel treatments	
	Season	Land use and land cover (deforestation, agriculture, plantations)	

(*continued*)

Table 4. Continued

Fire variable	Natural influences	Human influences	Fire regime parameters
Fuel moisture	Season	Climate change	
	Antecedent precipitation	Land management (logging, grazing, patch burning)	
	Relative humidity		
	Air temperature	Vegetation type and structure (species composition, cover, stem density)	
	Soil moisture		
Ignition	Lightning	Human population size	Number and spatial and temporal patterns of fires
	Volcanoes	Land management	
	Season	Road networks Arson	
		Time of day	
		Season	
		Weather conditions	

Source: Bowman, D. J. M. S., Balch, J., Artaxo, P., Bond, W. J., Cochrane, M. A., D'Antonio, C. M., DeFries, R., Johnston, F. H., Keeley, J. E., Krawchuk, M. A., Kull, C. A., Mack, M., Moritz, M. A., Pyne, S. J., Roos, C. I., Scott, A. C., Sodhi, N. S., and Swetnam, T. W., 2011. The human dimension of fire regimes on Earth. *Journal of Biogeography* 38, 2223–36. Table 1. Blackwell Publishing Ltd.

The construction of physical fire-breaks may be involved, or the use of fire itself in a controlled manner to reduce the fuel load in advance of a fire. Yet in spite of all these human efforts, it may take a change in weather to extinguish a fire, as in the case of the Yellowstone fires of 1988.

One of the important decisions must be what to do when fire threatens lives and property. This brings us to the important issue of building in flammable ecosystems and the wildland–urban interface.

Wildfires and urban populations

The attitude to fires in wildland or rural settings can be quite different depending on culture, history, and experience. Within some wildland areas fires have often been seen as part of the natural environment, and in some places human populations have regularly used fire as part of living in such areas. This can be seen historically in some areas of the south-western USA among native Americans or in Australia with the so-called 'fire-stick' culture of the Aboriginal peoples. Even in developed areas such as in Europe fire historically has played an important role in many forms of agriculture. In many places then fire has been seen as an important part of living with nature. In more urban centres the destructive force of fire is paramount and fire in such circumstances is an enemy and rarely a friend unless contained. This contrast between experience and attitudes has never been more stark than over the past century.

In a series of books discussing the modern history of wildfire Stephen J. Pyne of Arizona State University has coined the term 'Pyric Transition' for the change in attitude to fire that has followed industrialization. Fire had for centuries been considered as part of the natural system and was manipulated by those living in rural communities. However, a major change occurred with ideas about fire as population grew and urban settlements spread. There came to be a divide between countryside populations experiencing fire as part of a natural landscape and a useful tool in agricultural land management, and urban populations for whom fire became 'boxed up' to provide heat and energy for the population. This has caused increasing tensions at the wildland–urban interface (more appropriately termed the rural–urban interface such as in England). So now we are seeing a change in the narrative. Traditional ideas of landscape or pastoral fire and indeed agricultural fires are increasingly misunderstood by those living in urban settlements, where fire must be excluded (Table 5).

This change in our attitude to fire has many more consequences than can be seen at first sight. The spread of views about the

Table 5. Changing fire regimes before and after the industrial era

Biome	Pre-industrial fire regime	Post-industrial fire regime
Tropical rain forest	Very infrequent low-intensity surface fires with negligible long-term effects on biodiversity	Frequent surface fires associated with forest clearance causing a switch to flammable grassland or agricultural fields
Tropical savanna	Frequent fires in dry season causing spatial heterogeneity in tree density	Reduced fire due to heavy grazing causing increased woody species recruitment
Mid-latitude desert	Infrequent fires following wet periods that enable fuel build-up	Frequent fires due to the introduction of alien flammable grasses
Mid-latitude North American seasonally dry forests	Frequent low-intensity surface fires limiting recruitment of trees	Fire suppression causing high densities of juveniles and infrequent high-intensity crown fires
Boreal forest	Infrequent high-intensity crown fires causing replacement of entire forest stands	Increased high-intensity wildfires associated with global warming causing loss of soil carbon and switch to treeless vegetation

Source: Bowman, D. J. M. S., Balch, J., Artaxo, P., Bond, W. J., Cochrane, M. A., D'Antonio, C. M., DeFries, R., Johnston, F. H., Keeley, J. E., Krawchuk, M. A., Kull, C. A., Mack, M., Moritz, M. A., Pyne, S. J., Roos, C. I., Scott, A. C., Sodhi, N. S., and Swetnam, T. W., 2011. The human dimension of fire regimes on Earth. *Journal of Biogeography* 38, 2223–36. Table 2. Blackwell Publishing Ltd.

Fire

importance of fire suppression and the restriction of prescribed burning may have several unintended consequences. The change in attitude has been taken from the towns and cities back into the countryside by a population that has lost its connection to fire.

A problem has arisen when people have moved into wilder, often forested areas, without a full understanding of how fire occurs and behaves. Such a move is understandable: more than ever we wish to be surrounded by nature, and indeed we are encouraged to plant trees within urban settings. But even those who live within a flammable ecosystem such as parts of California and Australia may not be fully aware of how wildfire works. There are several aspects of which we need to be aware.

Wildfire may be an ever-present reality and danger in some regions. This may even be the case if there has been a recent fire as downed dead fuels may accumulate and invasive grasses may spread in the aftermath. In addition, we face challenges through climate change (irrespective of the cause). A small increase in periods without rain or an earlier spring snow melt may create additional problems.

And the speed of wildfires should not be underestimated. Fires may spread extremely rapidly especially if driven by wind. Fires may also spot, and new fires may start many kilometres from the main fire front. Satellite or space station images can show this quite dramatically.

Planning for escape in fire-prone areas is important, and should not be limited to a single road or direction of travel. The danger of lack of planning and awareness has been seen in many recent fires such as the Fort McMurray Fire in Canada and fires in Portugal where people have been killed on the road while trying to escape. The speed of movement of fires has surprised many, meaning that they leave it too late to escape. Some who stay to fight a fire to protect their homes may also underestimate the nature of a wildfire (Box 4).

In many flammable areas, buildings are constructed with flammable materials, and trees are planted close by. Even when fire-retardant building materials are used, fire may come on the wind through ventilation ports. These problems have led to the development of a movement called 'Firewise' in America and 'FireSmart' in Canada (and it is spreading to other countries) that alerts people and communities to the problems of building and living among flammable vegetation.

Even if people believe that they do not live in flammable vegetation in a fire-prone climate, climate change may create a sudden switch to more frequent fires in an area of currently little fire. Such may be the case in southern England where Surrey, to the south of London, is the most wooded county in England and where an extensive forest fire would be catastrophic. Even a small fire may cause significant problems if it is in a high-risk area.

Another problem that occurs when urban populations with an urban view of fire move into the countryside is that all wildfire is considered bad and to be extinguished. Fire suppression becomes the new normal and prescribed burning is not to be tolerated. In some circumstances this approach may lead to the build-up of fuel so that when a fire eventually erupts it is much more severe than it would have been. Our problem is that in a changing world our attitude to fire also needs to change.

Building and urban fires

What is not in question is the need to exclude unconstrained fire from buildings (Figure 22). In the past building fires were an ever-present danger. Buildings were made predominantly of

Box 4. How to survive a fire

While there are some similarities between urban and building fires and wild or landscape fires there are also many differences.

Most fires, except smouldering fires, have flames, but in both urban and rural settings these appear not to be the main causes of death. Three significant aspects of fire are smoke, heat, and gases. In contrast to flames, smoke, heat, and gases are a particular problem in buildings, as they are often confined within a small space, while they are less concentrated in landscape fires. Oxygen fuels the fire so although levels may be reduced it is usually rapidly replenished. Asphyxiation may, however, be a problem in many fire environments. More problematic may be the inhalation of superheated gases in both types of fire. Whereas smoke in landscape fires does not contain toxic compounds this may be a significant factor in building fires, in which household materials and plastics are combusted.

Early warning systems such as fire and smoke alarms and sprinkler systems may protect people from serious injury or death within an urban setting, but no such warnings or automatic suppression systems are available in rural situations where people will need to rely on experience, sense, knowledge, and skills to provide an early warning of a possible threat to life.

Surviving a fire within an urban or building context will depend much upon a risk analysis and prevention measures that are in place but any occupant must make themselves aware of all the precautions and procedures in place. In all circumstances remaining calm is important and this will help in the decision making needed. It is important in such a situation to acknowledge the stress you may be feeling; fear is only natural.

(Continued)

Box 4. Continued

Marty Alexander of the Canadian Forest Service has provided over the years a number of guides on how to survive a fire, concentrating on wildfires. There may be an option to escape from a threatening wildfire, but this may not always be possible. There are four simple concepts that may help reduce the possibility of being killed that should be followed at all times.

Perhaps the most important advice, surprisingly, is to remain calm and not panic. The second is to select an area that will not burn and the bigger the area the better. Certainly it should be an area with the least amount of combustible material, and one such is a depression, which offers the best microclimate. The third is to protect yourself from radiant and convective heat emitted by the flames, using the shield of boulders, rock outcrops, downed logs, etc. And the fourth point is to protect your airways from heat at all costs and to try to minimize smoke exposure. It is important to avoid dense smoke, perhaps by keeping low, and a damp cloth over the face will mitigate the heat being taken into the lungs.

There are four basic or fundamental survival techniques recommended for those caught in the open during a wildfire event, when it is not possible to take refuge. The first is to retreat from the fire and reach a place of safety. It is better to go downhill rather than up. The second is to burn out a safety area. The third is to 'hunker down' in place, and surprisingly, the fourth is to pass through the fire edge into a burned out area. This last may seem the most frightening but in some cases may offer a better chance of survival than trying to outrun a fire.

flammable materials such as wood, thatch, and even paper, for example in Japan. And they were lit and heated with open fires and candles. Fires could easily arise from knocking over a burning candle. And once started, there was for a long time no systematic approach to extinguishing a fire.

Box 5. Room fire

The way in which a fire takes hold within a room illustrates the challenges of extinguishing a fire and preventing it from spreading quickly to become a fully-fledged building fire. Consider a typical living room with furniture. A fire may ignite from various sources, most commonly an electrical fault, for example in a television, or from a naked flame such as a cigarette or decorative candle. In many societies there are regulations concerning the flammability of various furniture items but this does not mean that they will not catch fire. Suppose a chair placed near a wall is set alight. This can easily happen from, for example, a dropped cigarette.

Following ignition, the fire will first grow on one object (perhaps smouldering for a time) but may quickly spread to others nearby. In a normal room the fire behaves much like a wildfire in open air, but the emissions of smoke and carbon dioxide may be confined within the enclosed space, and smoke is often the killer. The burning of furniture increases heat intensity, and the accumulation of fuel-rich gases, which may come into sudden contact with the air, can create a backdraught. Under such conditions, the result can be what is called a flashover. The burning rate of the fuel increases rapidly, resulting in a sudden expansion of flames throughout the room.

If there is little ventilation in the room, especially if a fire door remains shut, and fuel is limited, the fire may peter out. Opening a door or window will allow a rapid ingress of air (oxygen) that will help the fire grow.

The time needed for a fire to develop can vary considerably. A smouldering fire may take more than twenty minutes before it changes to a flaming fire, after which the temperature of the fire may rise from 260 °C to 540 °C in a very short time. It is about this time that a sudden change in the nature of the fire, the

(*Continued*)

Box 5. Continued

flashover described above, occurs. Fully developed room fires may reach temperatures of over 1000 °C.

If the room has a window or door open, then smoke begins to collect on the ceiling before flowing down towards the floor of the room, in some cases being only a few centimetres above it. The variations in how fire moves both within and out of a room are complex and not completely understood, and this makes both building design and fire proofing and fighting difficult. What, however, needs to be appreciated is just how quickly such fires can develop, and the measures needed to combat fires and escape from them.

The result was that in population centres fires could easily spread from building to building, leading to major catastrophes such as the Great Fire of Rome in 64 CE, the Great Fire of London in 1666, and the Chicago Fire of 1871. This ever-present danger has led to many advances in techniques of fire suppression. One advance has been a change in the building materials used, and the introduction of fire doors, etc. However, despite many improvements fires always remain a significant hazard. A second advance was the enclosure of naked flames—a move away from open fires for heating and cooking and more importantly naked flame candles used for lighting (although a fire risk remains today even in first world countries where fire prevention is considered a priority, in the form of decorative candles and of course from the continued use of cigarettes). However, in some parts of the world both indoor open-flame cooking and the use of candles for light are still widely practised. Despite the reduction in the use of open flames within buildings fires may often be started by electrical faults or from gas explosions or careless use of chemicals, or even from cooking fat that has caught alight. Even though a number of materials are now fireproofed, many house fires still occur (Box 5).

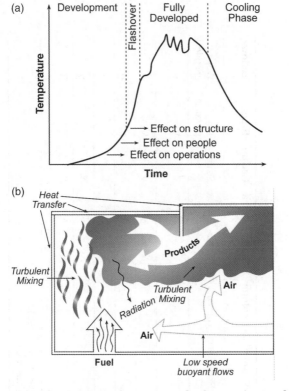

22. **Building fire.** (a) **Time and temperature development in room fire;**
(b) **Elements of a room fire.**

Another advance has been in fire prevention and monitoring.
Many buildings now have smoke detectors to give early warning
of a fire and fire alarms that can be set off automatically or by
human intervention. Warning of a fire within a building elicits a
response. The first responder may use a number of interventions
such as water hydrants, foam hydrants, fire blankets, sand
buckets, or carbon dioxide cylinders, depending on the nature
of the fire (Table 6).

Table 6. Fire extinguishers and use

Type of fire	Material	Red Water	Black CO$_2$	Cream Foam	Blue Dry powder	Yellow Wet chemical
Class A	Combustible materials (e.g. paper and wood)	√	X	√	√	√
Class B	Flammable liquids	X	√	√	√	X
Class C	Flammable gases	X	X	X	√	X
Class D	Flammable metals	X	X	X	√	X
Class E	Electrical	X	√	X	√	√
Class F	Deep fat fryers (e.g. pans with hot oil)	X	X	X	X	X

Green (Halogen) is no longer legal

Automatic water sprinklers may also be fitted and automatic fire doors may be included in building design, as well as clearly signposted escape routes. Despite this a fire may take hold quickly and require the intervention of a fire brigade that can offer a larger-scale suppression of the fire with specialist water hoses and other equipment. But even this may not be enough.

Building design and the use of certain materials may inadvertently contribute to the fire becoming more serious and create problems for its suppression. This has been seen especially in some high-rise fires in recent years. At the Grenfell Tower Fire in London in 2017, the fire spread extremely quickly, particularly via the cladding on the outside of the building. Such tragedies bring to public view failures in design, testing, and regulations, as well as unforeseen issues, and raise many questions concerning aspects of fire safety and fire suppression.

Fire, smoke, and human health

In many building fires fatalities come from smoke rather than the flames. Within a building environment smoke may be the first sign of a fire and a smoke alarm becomes an important part of a fire warning system. Smoke has four features that make it dangerous. The first is that it replaces oxygen in the air and makes breathing difficult. Secondly, because of its density, it may replace the air near the floor and may easily get into a ventilating system and spread quickly through a building. Thirdly, smoke seriously reduces visibility, making escape from a fire more difficult. The smoke from a building fire may also spread to areas around the building, causing further problems. Fourthly, toxic smoke may be emitted from the combustion of furniture and plastics that also causes choking and oxygen starvation and may be a hazard not only for any inhabitants of a building but also for fire fighters (Box 6).

Box 6. Smoke and atmospheric pollution

One of the long distance signs that there is a fire is from the plume of smoke that is given off and is seen rising into the atmosphere. The amount of smoke and its colour help identify the condition and nature of the fuel being burned. Whether the smoke is white, grey, or black will depend on the amount of water vapour but also on other materials. These include small charcoal particles, usually less than 125 µm in size if dealing with vegetation fires, but sometimes larger particles will be lofted in the plume. One significant element may be soot, or black carbon. Soot is formed by the recombination of vaporized organic molecules to form a new substance that is almost pure carbon. Soot consists of particles less than 1 µm in size and their shape depends on what has been burned, so soot from petrol and soot from vegetation burns have different shapes. It is this soot that causes blackening of snow in the aftermath of a fire. A range of gases and aerosols may be incorporated into the smoke plume, including carbon dioxide, carbon monoxide, methane, the oxides of nitrogen. In addition there may be other organic compounds present, such as polycyclic aromatic hydrocarbons (PAHs), and again their composition may vary depending on the material that is being burned. Because of this the nature and origin of a fire can be determined from the chemistry of the smoke plume released.

The smoke from a fire can stay in the atmosphere for a considerable period and may have the effect of preventing the formation of raindrops. The nitrous oxides in the smoke are particularly hazardous to human health. Smoke plumes may be seen and mapped by satellites. The plumes can easily reach heights of 5 km and have been seen spreading more than 1,000 km from their origin as the smoke can be carried into the troposphere by high altitude winds. Carbon monoxide from wildfires in Alaska has been shown to spread right around the globe.

While smoke from building fires has long been known to be dangerous, the smoke from wildfire has only recently been identified as a serious health hazard. Studies on the health impact of wildfires around Darwin in northern Australia by Fay Johnson of the University of Tasmania showed significant health impacts for those who were regularly exposed to smoke. We may be familiar with television images of smoke pollution from Indonesian peat fires (Figure 23) spreading not only to nearby Singapore, as experienced in July 2013, but also as far away as Kuala Lumpur in Malaysia. Such smoke plumes from fires can now be seen both from satellite and space station images, and this new data allowed Fay Johnson and her colleagues to correlate and plot deaths from wildfire smoke globally in 2012 for the first time.

One of the key aspects of smoke that affects health is the presence of very small particles (known as PM10), less than 10 µm in size. These get into the lungs and can cause lung disease. The source of the particles can be identified using microscopic techniques.

23. Smoke from peat fires in Indonesia.

The health impacts of smoke are not limited to breathing problems. Studies have shown that if pregnant women are subjected to significant smoke from peat burning (as they have been for example in Indonesia) during the early phases of pregnancy, they are more likely to suffer a miscarriage or give birth to children with deformities.

The fear of smoke is a key reason why some agricultural burning techniques and surface fuel burning have been substantially reduced, but this may in turn lead to other problems.

Chapter 5
New technologies and changing fire policies

Two factors have led to a fundamental change in our understanding of fire as an Earth System Process. The first is our growing knowledge of fire in deep time before the evolution of humans, which demonstrates that fire has played a significant role in shaping the Earth that we know. The second factor has been the new ways we can observe fire today, especially the rapid progress in satellite monitoring of fire over the past 20+ years.

Observing fire

For much of human history we have relied solely on our eyes to see and report fire. Even then, there were not often records of a wildfire or even of most building fires. The result of this was that we lacked an understanding of just how common wildfires were across the world. If we had little knowledge of wildfire occurrence and extent in the modern world we had even less understanding of fire through deep time, the hundreds of millions of years before humans when fires would have occurred.

An increase in our knowledge came first with the rise of communication, when news of a fire could be transmitted across large distances. Such was the case in 1910 in the USA, in what has been called by Stephen Pyne the 'year of the fires'. This was a catastrophic year for wildfires, in part because of a dry summer. But it also shaped how the public and especially politicians viewed fire, and particularly its suppression. The events of 1910 altered attitudes to wildfire considerably, with significant consequences over the subsequent decades. Preventing fires and putting out all fires became the mantra, and such was the influence of the United States that such policies were adopted across many other countries and continents.

There are many parts of the world in which fires occur that are not witnessed or reported by humans. The failure to keep records of fire has also been an issue especially when trying to develop an understanding of fire size and frequency. Even until the 1960s much of our observation of wildfire came from airports, where smoke affected the takeoff and landing of aircraft from population centres where fire threatened habitation.

All this changed with the space race. We tend to forget that it is really quite recently that our first satellites were put into orbit. But even then these did not produce publicly available images of the Earth. In the 1970s, the first Landsat images of large areas of the Earth became available. Not only could images be seen in ordinary white light, in which for example smoke plumes from burning fires were visible, but information could also be gathered by mapping in wavelengths of light beyond the visible that allowed living and dead vegetation to be viewed in different colours. For the first time this enabled the sizes of wildfires to be estimated from studies of the burned area, which would have been difficult to do on the ground because of topography and access. The analysis of such landscape images became ever more sophisticated through the 1970s and 1980s with, for example, Burned Area Emergency Response (BAER), which were maps of burn severity, being produced in the United States of America. These not only produced maps of the area affected by a wildfire, but allowed a much greater understanding of the impact of a wildfire across a region.

Increasing numbers of satellites were launched in the subsequent decades particularly by the United States but also increasingly by Europe, and these satellites had two main advantages. First, they were placed into different orbits. Some were placed in a fixed orbit, remaining above the same spot on Earth, so that it could be observed continuously over a twenty-four-hour period. Others followed a polar orbit that could scan the whole Earth and allowed the production of maps of fire across the world over a twenty-four-hour period. The second advantage for the monitoring of wildfire came from the instruments that such satellites carry.

In the 1980s the Advanced Very High Resolution Radiometer (AVHRR) was developed, which could scan the Earth's surface using a number of different wavelength channels. These could not only look at the occurrence of fire, using, for example, thermal infrared data from which temperatures could be derived, but also determine the size and nature of smoke plumes from the fires. Over the years the resolution of the instruments has increased so that now data can be derived from remote sensing imaging at a resolution of 1 km. Much of the data from satellites has been collected for other purposes but has proved useful for wildfire research. One area of interest was to develop sensors to monitor active fires both in the day and the night that would allow us to determine how fire relates to climate as well as human activity.

A particularly important development was the launch in 2002 of the European Space Agency satellite called ENVISAT that was used to monitor environmental and climate change. The data derived from this satellite were used to develop a global fire atlas, in which each year requires data from the processing of 80,000 images with hot spots with a temperature higher than 321 °K at night, precisely located to less than 1 km. This has revolutionized our understanding of wildfire.

Likewise we are all now familiar with views of fires from the international space station, usually recognizable by extensive smoke plumes billowing across a region. Other developments also included products to classify burned areas and these data have proved useful not only in assessing damage to the environment but also in allowing calculations of carbon loss and carbon transfer in the environment.

There is no doubt, however, that the most significant development in instrumentation of value for fire detection over the past twenty years has been that of the MODIS product (and Fire Radiative Energy estimations). The MODIS instrument, or Moderate Resolution Imaging Spectra-Radiometer, is aboard the Terra satellite that was launched by NASA in 1999. The large amount of data obtained by the instrument since 2000 can be used in a variety of ways; so, for example, the Fire Information for Resource Management System (FIRMS) puts together both remote sensing and global positioning systems technologies to deliver both global MODIS hotspot fire locations as well as burned area information and builds upon a mapping interface to give near real-time hotspot fire information and monthly burned area information. At least some of these data are now readily available using 'Google Earth'.

What has been produced is truly remarkable and has led to substantial improvements in our understanding of the distribution of fire on Earth. It also presents many new challenges for policy makers. This is because it shows the number and size of fires throughout the world on a global scale, and confirms that in many regions fire is an important part of the ecosystem. This is well illustrated by the occurrence of fire in Africa through the year where early in the year human-ignited fires predominate in central Africa, yet the band of fire spreads southwards through the year where natural fires occur in savanna and fynbos. We can also see fires in relation to political boundaries that indicate different approaches to fire suppression and human ignitions. What becomes clear is that fire appears to be a natural part of the ecosystem in some regions but not in others. This is especially evident from observing the occurrence of fire in the Amazon rainforest, where almost all the fire is related to human activity.

After the fire

Let us now move our focus from the global to the local, and indeed to the very essence of the interaction of a surface wildfire with the soil. Until now our concentration has been on the fire itself, and while new maps can be produced of area burned, the significance of the aftermath of a fire is often overlooked. We have seen that fires often begin as surface fires and our attention is focused on the flames and later movement of a fire, in some cases its upward movement from the surface to the crowns of trees. Yet we must not forget that the heat of the fire may also be transmitted down into the litter layer (the layer of dead organic matter lying on top of the soil) and indeed into the soil itself.

Fire can have several effects on the soil. The slow-moving surface fire may simply char the litter layer as the flames pass, as not all of the vegetation may be fully consumed by a fire. However, the build-up of extensive litter fuels, either naturally or as slash, may allow a fire to burn more intensely and at a higher temperature. This heat may also affect the top layers of the soil, destroying the roots of the plants, which often bind the mineral soil. Root systems may continue to be subjected to continued downward heating long after the fire front has passed. As we have already seen, the charring process produces volatile gases and liquids that may be released upwards into the air, but also downwards, to be precipitated within the soil itself.

The effects of fire on soils are particularly significant. Many soils may develop a water repellent layer. This is especially the case if the soil is organic rich. How many of us may have potted a house plant in an organic-rich soil only to find that when we water, the liquid simply pours off the surface and takes a considerable time to penetrate down into the soil? We know that many soils, especially those under forest and those associated with certain types of vegetation such as chaparral scrub (as in California) and *Eucalyptus* forests (particularly in Australia, but also now widely planted elsewhere, such as in Portugal), have highly water repellent soil layers. These water repellent layers are underlain by a wettable layer, usually a mineral rich soil. However, a surface fire can dramatically alter this simple structure, so that the positions of the wettable layers and water repellent layers are shifted. In some cases the uppermost water repellent layer may be enhanced, as happens in some coniferous forests. In the case of chaparral scrub the water repellent layer may move, giving a sandwich of an upper and lower wettable layer with an enhanced repellent layer between. This is taken to an extreme in *Eucalyptus* forest, with the repellent layer becoming almost impenetrable.

Post-fire erosion and flooding

What is the significance of this? We need to consider now not only the drying of a soil but also how it responds to rainfall. On a slope, even a small one, rain predominantly will soak into and penetrate a soil. In exceptionally heavy rain some of the water may move over the soil by what is known as overland flow. We may see this in our own area or regions or even if we are unlucky in our gardens. It is the reason why there is such concern about tarmacking or concreting over large areas in cities; the result is an increase in overland flow and flooding. However, we have noted that the fire may not only have altered the soil structure by introducing an intensely water repellent (hydrophobic) soil layer but may have also destroyed the binding roots of plants. In such a case if there is intense rainfall following a fire then the upper, shallow, wettable (hydrophilic) layer may become rapidly saturated and when the water reaches the impenetrable layer below, it will begin to move by overland flow, bringing with it large amounts of sediment and in many cases charcoal from partially burned plants. The destruction of the surface vegetation can mean that the flow may move extremely rapidly and a slurry of water and sediment rushing away from a burned area can cause massive destruction many kilometres from the burn site.

Where there are steep slopes, extensive channels may be cut and sediment moved in large quantities across the landscape, which can result, even overnight, in the deposition of large bodies of poorly sorted sediment such as alluvial fans (Figure 24). It is only in the past few years that scientists have appreciated the significance of post-fire erosion and the hazards it poses. And the problem only became widely reported in the past couple of years, following the inundation of homes of Hollywood celebrities with flood waters and sediment following major wildfires in California.

The floodwater may carry a range of insoluble and soluble materials including nitrates and phosphorus that may have an impact on stream, river, lake, and reservoir water chemistry. When drinking water is contaminated this adds a further problem. This happened in the case of the Hayman Fire in 2002 in Colorado, following which the main water supply to the city of Denver was badly affected.

24. Post fire deposition of alluvial fan.

We may consider fire suppression but how do we prevent post-fire erosion? Strangely enough the push for finding solutions has come from the insurance industry, which has had to pay out for flood damage, often a long way from a wildfire. Many attempts have been trialled, including felling trees across the burned slope to check the movement of water and sediment, the putting up of artificial barriers, and the dropping of hay bales across burned areas. These burst when they hit the ground providing a layer of organic material that can absorb the first rainfall and prevent the initial movement, albeit only in those areas treated. The problem is that burned areas may be large and the time needed to treat may be too long. In addition, in some cases exotic plants, especially grasses, have inadvertently been introduced with the hay, which creates problems later. Much research is still needed but the first step has been to recognize that there is such a problem.

Fire and legislators

As more of the world's increasing population moves from the countryside and into cities, and fire prevention becomes a pervasive issue, we develop the attitude that all fires are bad and must be suppressed. Before considering fire within an urban setting we should first think about fire in the countryside and in wildland areas. As we have already noted, the 'Year of Fires' in 1910 in the USA led not only to profound changes in the attitude to fire but also to legislation surrounding fire prevention and suppression. Despite the knowledge built up by local farmers, agriculturalists, and native populations who may have successfully lived with fire, the threat of large wildfires both to the economics of the forestry industry and to expanding urban populations made legislation inevitable. The 'Smokey Bear' campaign in the United States has become the symbol for preventing forest fires and this has led to other similar successful campaigns in other parts of the world, such as Spain (the All Against Fire (Todos Contra el Fuego) campaign in the 1980s and 1990s). Even today there is still a pervasive attitude that all fire is bad. This leads to a number of problems and conflicts.

We have already seen that fire may be a natural part of one vegetation type in a region but not part of another. For example, fire is an integral part of savanna grasslands yet not a natural part of many tropical rainforests. What do we do when such different ecological systems come in to contact with each other, especially within a single country?

Ideas about fire and about the origin of some types of vegetation can also be incorrect. It has often been asserted that many grasslands in Africa (including the island of Madagascar) are recent and are only present as a result of the degradation of wooded areas. Research by William Bond and Sally Archibald and others from South Africa has clearly shown that the grasslands are highly flammable, accounting for as much as 70 per cent of the world's burned area, yet the climate is also wet enough to support closed forest. However, it has been shown that many of these grasslands are truly old and not a result of humans creating them by large-scale deforestation. If that is the case then it is highly significant.

If we consider an area such as the island of Madagascar, which is a world biodiversity hotspot, all attention has been paid to its tropical forests with its attendant unique plants and animals. This has led to legislation based upon the idea that all fire is 'bad' and should be prevented and extinguished. Yet what is not generally appreciated is that grassland areas are also highly biodiverse, as shown in the 2017 'State of the World's Plants' review by scientists at the Royal Botanical Gardens in Kew. In addition, we now know that these grasslands are themselves ancient and existed before human arrival and are not, therefore, a consequence solely of human activity. Such biodiverse grassland can only be maintained by fire. What then is the solution?

Perhaps the first part of the solution is to raise the role of wildfire in the Earth System among scientists themselves. At the launch of the Kew Report in 2017 I was invited to give a talk on Fire in the Earth System (as was William Bond). What surprised me most was that many conservationists had not appreciated the role of fire in a number of ecosystems. Secondly, we need to share our attitude towards and understanding of wildland fire with the general public—only then may it come to the attention of legislators. We need unbiased debates, but when discussions begin a conflict develops between different interest groups. For example, land managers of heathlands may argue for prescribed patchwork burning, but others may object to such a practice either from their own conservation perspective or on grounds of smoke pollution. Unfortunately such debates can become very polarized, especially when the media becomes involved. This then is a problem, as we need the media to help educate but the focus is more often on the reporting of conflict.

Scientists have sought to publicize the issues through articles and books, and propose ways ahead, as with the Chicheley Declaration, 'A Vision for Wildfire Research in 2050', signed by a group of scientists meeting at the Royal Society in 2015. However, the issue of fire on the landscape may seem less significant than issues around flooding and flood prevention. Yet global climate change may alter that. In the UK the Parliamentary Office for Science and Technology in 2019 produced a POST Note on 'Wildfire and Climate Change' that provides a summary for all members of the UK Parliament to help them in making policy decisions concerning wildfire. This is an important development and may provide a useful template for other countries to follow.

Fire safety and suppression in the urban setting

It is important to consider how we can live with fire and develop a sustainable coexistence. There are different considerations in both the rural and urban landscapes, but these come into contact with each other so we need to be aware of both situations. We need to consider when and how and even if we should attempt to tackle a fire. Fire suppression and prevention within urban centres is much more focused and prioritized, but as many tragedies have illustrated, and continue to do so, there is much progress to be made.

The importance of the wildland–urban interface (WUI, or rural–urban interface) has recently become more appreciated. We can look at this interface in two ways. Perhaps the most obvious is that we are building more into flammable vegetation so that the threat of fire becomes more apparent. In 2009 the Black Saturday fire in Australia caused a large number of fatalities, 173 in the Kinlake area. To what extent can we plan or legislate for this activity? Even if there are plans in place the irregularity of wildfire may make a population complacent. In 2018 in Greece in the coastal area of Attica 102 people were killed trying to escape a wildfire as some planned escape routes had been cut off and strong winds caused the fire to spread quickly. A number of people were found dead huddled on the beach.

The devastating Camp Fire that destroyed the town of Paradise in north California at the end of 2018 illustrates the point, Here too, there was much loss of life—at least eighty-five dead. In this example—and there are many more—the town comprised mainly wooden framed buildings and structures, many of the houses were surrounded by trees, and the number of routes out of the town was limited. This is a situation that exists in many parts of the world. Could a better understanding of fire among the inhabitants and legislators have helped, with better preparedness for a wildfire, changes to building regulations, and indeed regulations within the urban context? Such considerations have led to the idea of defensible space, involving good practice concerning the construction of the house and also the vegetation surrounding the dwelling. Even this, however, is no guarantee of safety.

The next step in the United States, an approach that has been taken up in other countries, is that of the development of 'Firewise' communities and community-owned solutions that includes a significant input of public education.

There are a number of firewise precautions that are often recommended by a range of government agencies in the United States but many of these can easily be adapted for use in other countries (see Box 7 and Figure 25).

Yet all this advice may not be enough; legislation may be required not just on a local or regional level but also on a national level. Legislation concerning the building of new towns in England, for example, requires consideration and planning for fire at the wildland–urban interface, with planning for routes for fire suppression vehicles.

Within buildings themselves we need to think about three important issues. Most important is the flammability of building materials. Legislation requires they meet flammability requirements, but as has been shown recently following the Grenfell Tower Fire in London in 2017, the small-scale tests may not indicate what may happen in a large building. The experimental testing of materials almost certainly needs to be linked to computer models that look at how materials may behave in a range of scenarios, and lessons learned about how fires spread. This then is part of fire prevention. But fires cannot be completely excluded from buildings, because they result mainly from faults within electrical equipment.

We need in the first instance to have an early warning of a fire. Good fire and smoke alarm systems, therefore, are essential and although public buildings require them some have significant limitations, and many private residences do not have any such systems. When a fire starts and an alarm sounds, the nature and speed of response may be critical.

Box 7. Wildfire preparedness checklist

In an area surrounding a house there are a number of simple rules:

determine the size of an effective defensible space;

remove dead vegetation;

create a separation between trees and shrubs;

create a separation between tree branches and lower growing plants;

create a Lean, Clean, and Green Area extending at least 10 metres (30 feet) from the house;

and finally, maintain the Defensible Space Zone (Figure 25).

Consider the plants, shrubs, and trees to be grown. Especially avoid planting *Eucalyptus* and some pine species and certainly not near a house.

Many checklists for home owners include the following:

replace wooden shake and shingle roofs with fire resistant types such as composition, metal, or tile;

plug roof openings and gaps in roofs with non-combustible materials;

keep roofs cleared of combustible debris such as pine needles and leaves;

replace plastic skylights with double pane types in which one layer is tempered glass;

install an approved spark arrester on chimneys or stove pipes;

replace single pane windows with multi-pane tempered glass types, and close windows if wildfire is threatening;

(Continued)

Box 7. Continued

cover attic, crawl space, and eave vents with 1/8 wire mesh, and change plastic vent covers to heat-resistant metal covers;

keep rain gutters free of combustible debris;

fill gaps in siding and trim with good quality caulk, but also consider replacing siding with non-flammable construction materials;

store firewood and timber well away from home, 10 metres (30 feet) if possible;

place combustible patio furniture inside house or garage if wildfire is threatening;

replace deck boards that are in poor condition or less than one inch thick;

keep gaps in the boards, on top of the deck and the area under the deck free of flammable debris; clear decks of all flammable items like newspapers, wicker flower pots, dried plants, and gas grills when wildfires threaten, and use a lattice and metal mesh to screen under the deck to prevent the build-up of flammable debris;

keep flower boxes well maintained and free of dead material, and remove them if wildfire threatens;

box in eaves with plywood or other non-flammable construction material;

landscape flowerbeds with non-flammable plants, remove any dead materials, and replace any wood mulch next to house;

back vehicle into garage with windows up or park away from house when wildfire threatens;

adjust garage doors or use trim to have as tight a gap as possible to minimize ember entry, and close door when wildfire threatens;

cover rubbish bins with lid and move away from improvements;

keep wooden fences well maintained and utilize a non-combustible section or gate within 1.75 metres (5 ft) of the house or garage.

Stopping a fire early in its development is obviously critical. All public buildings are in most countries required to supply some sort of initial response to a fire via fire blankets, sand buckets, fire extinguishers, or specialist fire hydrants. Even in a house, keeping a CO_2 cylinder in the kitchen, for example, may be critical in preventing the spread of a fire.

Building fires may spread very rapidly and easily get out of control and require more specialist help in fire suppression. Even a single building fire may require the use of large numbers of fire fighters and specialist appliances for many hours. Yet even in the 21st century we still face many of the challenges faced by those dealing with fire as far back as 1666 in the Great Fire of London.

The first challenge is in how a fire may spread. This will depend on a number of factors that include the building materials, building construction, fire protection measures available (such as internal sprinkler system), the stopping of the spread of a fire, and the intensity of a fire. Each situation may require a range of control measures, from the use of specialist foams to the use of water delivery via helicopter rather than by the traditional fire engine. Even with a large number of fire engines and fire fighters many buildings may be almost completely consumed, but a key success may have been not in saving one particular building but

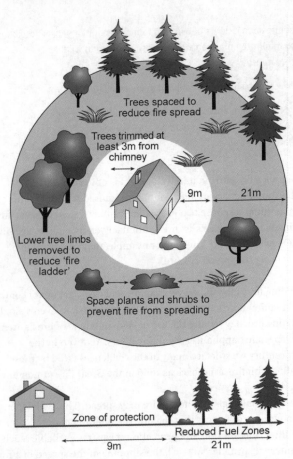

Trees spaced to
reduce fire spread

Trees trimmed at
least 3m from
chimney

9m   21m

Lower tree limbs
removed to
reduce 'fire
ladder'

Space plants and shrubs to
prevent fire from spreading

Zone of protection

Reduced Fuel Zones

9m   21m

25. **Firewise precautions.**

in preventing fire spread to adjacent buildings and especially
saving lives.

The second challenge relates to the contents of a building, as this
may have an impact on how the fire is fought. Clearly a factory

storage unit with paper will be treated differently from an empty building or one that holds flammable materials such as a petrol storage facility (e.g. Buncefield Fire, in an oil storage facility in England in 2005). Other buildings with important contents, such as Windsor Castle, one of the Queen's royal residences in England, which was damaged by fire in 1992, may require a different approach.

The third challenge is the aftermath of the fire. This may not only concern the structural integrity of a building but also a forensic analysis of how a fire started and spread. Important building fires may lead to a public inquiry to seek answers and provide future guidance for fire fighters and most particularly legislators.

The key to progress in this field comes from new knowledge in safety engineering and new ways to test fire resistance, not only of individual materials but also of whole structures. In addition there has been considerable progress in the design of fire vehicles. It is now widely appreciated that vehicles designed for tackling building fires have different requirements from those attending wildfires. Equally, the training of fire fighters to operate in these different environments may vary and specialist courses may be required. The development of fire suppressants has also played an important role. Within an urban context there may be ready availability of water but we also know that for some fires the use of foams may be required. Within a wildfire context water availability on the ground may be an issue, so much effort may be made by fire fighters to create fire-breaks and call in airborne vehicles carrying either water or fire-retardant liquids, gels, and foams that often contain organic solvents, foam stabilizers, and corrosion inhibitors. These cool the fire and cut off the supply of oxygen to the fire. New developments include producing additives that are effective with minimal water and coat the land surface so fire does not reappear.

Unusual fire

While we may be prepared for fire within most environments, there are some that provide special challenges. Within the natural environment we have already seen that peat fires may produce a significant challenge. This is because the organic peat acts as a fuel and will continue to smoulder long after the flaming fire front has passed. In addition, such smouldering fires can become very hot. If water is added to try to suppress a fire, then the water can evaporate before reaching the smouldering front, which may flare up again into a flaming fire. This has been the situation in the peat fires in Indonesia, and was also seen in the Saddleworth Moor fire in England in 2018, in which the fire moved from a surface fire through heather into the peat below, making extinguishing and controlling the fire much more difficult.

Underground coal fires may also present the same problem but can occur on a much larger scale and over a longer time period. The burning of coal underground can be made much worse by the release of flammable gases. However, it is the scale of such underground coal fires that can be quite staggering. Some of these may spontaneously combust because of heat generated by oxidation reactions, sometimes starting at the surface before spreading underground. A coal fire in the Indian coalfield of Jharia is over 280 sq. km and has been continuously burning for over 100 years, giving rise to major subsidence and severe health problems in the surrounding population through emissions from the fire.

Fires may be a problem in other fossil fuel deposits not only because of the ignition of natural gas or oilfield seeps, but also from accidental or deliberate ignition of exploration or production wells. In such cases fires may only be extinguished through the use of explosives that starve the fire of oxygen. Such fuel fires also occur in the air or on the ground through aviation accidents, but we tend to forget that other materials such as hydrogen are very flammable. The latter has resulted in fires such as that which famously destroyed the *Hindenburg* airship in 1937 and that of pressurized burning gas in the Space Shuttle launch of 1986. Surprisingly fires do occur in space, and care is needed even in the International Space Station as there are supplies of oxygen that may fuel a fire.

There are also fires that may occur in human constructions that require specialist suppression measures and create challenges to extinguish. These include fires within basements, where it may be difficult not only to suppress the fire but also to predict its spread. Also, fires within long tunnels, such as road or rail tunnels, bring many challenges. The fire in the Channel Tunnel linking Britain and France which occurred in 2011 caused the closure of the tunnel for several days, despite there being extensive prevention measures in place. The fire was initiated on a Eurotunnel shuttle that was carrying heavy goods lorries and lasted over sixteen hours with temperatures of over 1000 °C. Some tunnel fires such as the Mont Blanc road tunnel fire in 1999 have resulted in large numbers of fatalities.

Chapter 6
Fire and climate change

We often see pictures and reports of wildfires from across the world, especially if they are large and threaten habitation and, in some cases, cause fatalities. This was the case in November 2018 when there were a number of large wildfires in California (Figure 26). These were described as megafires and included the Camp Fire, north of San Francisco, which claimed at least eighty-five lives. These high-profile disasters generated extensive interest about why so many large fires should occur. A common question from the media to fire experts was whether the fires were so large because of climate change. Interest was also raised on this matter because the President of the United States appeared to deny climate change.

Climate change discussions revolve around three questions. The first is whether or not there is climate change; the second is, if the measured trends of changing climate are true, whether human activity is the cause; and the third, if both the others are true, is whether climate change can be reversed or stabilized. In this case, the relevant question appeared to be: is there climate change, and can this be affecting the frequency and size of wildfires?

There is general scientific consensus not only that climate change is happening but that human activity in the form of the release of greenhouse gases is accelerating the warming of the planet and

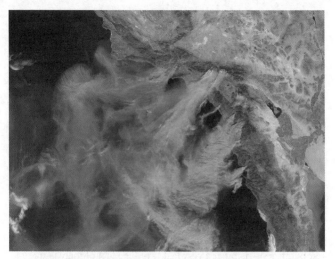

26. Satellite images of the Californian fires 2003.

creating unstable weather patterns. Even if we are able to stabilize the warming to the internationally agreed target of 1.5 °C by 2050 we still have the current problem of many areas of the world suffering from increased instability of the weather and a change in rainfall patterns that is affecting and will affect wildfire in the future. Engaging in arguments on the causes of climate change does not help the immediate and pressing issues concerning wildfire occurrence and impact.

The problem in answering the question is that it might become the single issue to be discussed; yet there are many interrelated issues. Indeed when I myself was asked the question concerning California wildfires, I started by stating that fires would be expected in this area with this type of vegetation. The second issue is related to building towns within such a flammable environment, especially where the construction of the houses is predominantly of wood and the trees are near many of the houses, a point often evident from television images. The third issue is that many fires have been made worse by the problem of invasive grasses; and the fourth issue relates to forest management and fire suppression.

Let us start, however, with the single issue of climate change irrespective of its cause. We can make three important observations. The first is that climate change is part of how the Earth works. There is what can be termed 'climatic variability'. The climate changes constantly over timescales of hundreds, thousands, and millions of years. So the relevant question is not whether the climate is changing but whether it is changing faster than would normally be expected. And if that is the case, how might that affect both the vegetation and also wildfire activity (Box 8)?

There is broad consensus among scientists that the global temperature is rising at a very fast rate. The fact of increasing global temperatures has widely, but not universally, been accepted by the population at large. What does this mean for our analysis? We should not confuse weather with climate. Climate reflects the overall state but weather concerns short-term variations. What climate change can do, however, is disturb established weather patterns. That may mean that some areas become colder or warmer for different parts of the year, rain may be distributed differently through the year, and more extremes of weather may be experienced.

The first impact of a warming climate is on the vegetation itself. Earlier and longer springs may alter how plants grow, flower, and fruit. Also some plants may suffer with even small changes in temperature, while others might spread from other environments, changing or enlarging their range. Animals can be affected too, especially insects. Some insect pests, often bringing fungal diseases, may be able to invade and spread, killing trees as they do so. Such is the case with some bark beetles in the western USA where many trees have died as a result of infection and this allows areas of dead plants to add to the dry fuel load. Moreover, if spring

Box 8. Fire and climate forcing

The term 'radiative forcing' is defined by the Intergovernmental Panel on Climate Change (IPCC) as the 'change in stratospherically adjusted radiative flux at the tropopause, compared to 1750 AD'. Essentially this refers to the change in heat as measured at the point where the troposphere (the basal layer of the atmosphere) meets the stratosphere above. Positive forcing will increase mean surface temperature, while negative forcing will cause it to decrease. Fires, then, can change radiative forcing through altered atmospheric composition (such as the release of CO_2) and changes in surface albedo, the proportion of light reflected by a surface. Burnt areas are darker, and soot may be deposited on snow, so the albedo goes down after a fire. The emissions from several types of fire have been calculated.

It was calculated by David Bowman and colleagues in 2009 that fires have contributed up to about 19 per cent of the anthropogenic (human-produced) radiative forcing since the pre-industrial era. The calculations are complex and based mainly on fire-related CO_2 emissions from wildfires (Figure 27). Fire also represents an important source of ozone precursors such as nitrogen dioxide (NO_2), especially in tropical regions. (Ozone is a less stable form of oxygen that in the upper atmosphere shields the Earth from harmful ultraviolet radiation. In the troposphere, nitrogen dioxide, not molecular oxygen, provides the primary source of the oxygen atoms required for ozone formation. Sunlight splits nitrogen dioxide into nitric oxide and an oxygen atom. A single oxygen atom then combines with an oxygen molecule to produce ozone.)

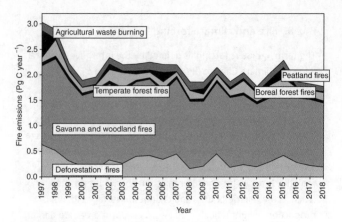

27. **Emissions from different types of fire.**

comes earlier in the year then the growing season increases, and by this means, too, a larger fuel load may accumulate.

The second aspect relates to the impact of longer and increased warmth on the landscape. This is particularly significant for the occurrence of wildfire in the Western United States. A warming climate means that there is an earlier snow-melt in the mountain regions. This in turn releases water into the system earlier. Following initial lush growth, there is then more time for vegetation to dry and hence fuel may become more flammable. In ecosystems that are already flammable and experience wildfire this may be particularly significant. Research by LeRoy Westerling from the Sierra Nevada Institute of the University of California has identified just such a small but significant change in spring temperatures.

The third impact is partly related: not only is the length of the potential fire season increased during the longer, drier summer but the weather may subtly change, with winds becoming more frequent and stronger. This would not only accelerate the drying of the fuel but also drive any fires once started.

So the answer to the first part of the question relating to climate change as seen in California is yes: climate change would appear to play a role in the changing nature of wildfire size and intensity. But we cannot just consider the question in relation to California—we need to think of climate changes elsewhere. An additional problem we face here is a very short-term memory and limited experience.

Various parts of the world are affected by the phenomenon of El Niño, and its opposite, La Niña. During an El Niño phase, parts of the western Pacific become particularly dry. It was during this phase in 1982/3 that major fires occurred. In contrast, more fires occur in the south-west of the United States during the La Niña years. The implication is that there are already cycles in fire activity as a result of these ocean-driven changes, but that does not take away from the observations of longer-term rises in global temperatures.

We also need to consider areas away from the western United States. In Australia, for instance, climate is distinctly changing, with longer drier periods, droughts, and a large number of extremely hot days. For vegetation that is already flammable, this shift can create significant problems. Even in England, not known for large wildfires, the number of wildfire incidents in recent years has been several factors higher than most people would imagine. Two hundred and sixty thousand vegetation fires were reported over an eight-year period, although only around 3 per cent of these would be considered true wildfires by the British Fire and Rescue Service. But that is over 1,000 per year, many more than most would imagine. This is only likely to get worse with increasing temperatures and an increase in the number of dry days. It has recently been shown that since 1884 the ten hottest years in the United Kingdom have all occurred after 2002.

I have already mentioned the risk of a serious wildfire in Surrey. Many small fires have occurred here, usually in heather heathlands. These are slow-moving fires and easily extinguished, and the vegetation quickly regenerates. If however the fire spreads into the surrounding forest (often coniferous), this adds several layers of complication. If the fire remains as a surface fire, even within the forest, it may move relatively slowly but access to extinguish the fire may be difficult. A good example was the Swinley fire in 2011 that occurred during a very dry period in the forest around Crowthorne on the Berkshire/Surrey/Hampshire border. It took over 400 fire fighters from eleven counties across England to get the fire under control. It was essential to prevent the fire from becoming a large crown fire as it might have spread more rapidly and into population centres and even industrial research facilities nearby. In addition, it was near major road arteries (M3 and M4 motorways) that would have been seriously impacted by the smoke and also near the flight path to Heathrow (London) airport. Smoke could have also contaminated nearby reservoirs. This fire has become a 'wake-up call' for those considering the impact of climate change in the United Kingdom.

If, then, areas of the world not familiar with wildfire may become affected, there will be few places in the world in which wildfire will not be a significant phenomenon as the century unfolds.

There is another consequence of climate change that needs to be considered. This relates to the migration of different plant species. Changes in rainfall and temperature may mean that the ranges of native vegetation may shift and this may have an impact on wildfire occurrence, type, and severity. However, the introduction of exotic (non-native) plant species by humans is also causing problems.

Plant invasives

One of the most surprising aspects of how landscape (wildland) fire has changed comes from growing recognition of the impact of the uncontrolled spread of certain exotic plants or what have been termed plant invasives. These are plants that have been introduced often from another country or region for one purpose and in some cases have spread across a region. In some cases a few of these exotic species have been deliberately introduced without an understanding of the fire hazard they cause.

Some of the grasses that have often been introduced for cattle feed form an important example. Often they have been selected for their rapid growth and high yield, resulting in a good grass crop for fodder. In the case of increased fire risk in the western United States, cheatgrass (*Bromus tectorum*) has been the culprit. Originating in Europe, Asia, and Africa, this grass is highly flammable, and not only has rapid growth and high seed production but also spreads very rapidly across a range of ecosystems. It may easily spread along roadsides and even grow in cleared forest areas. Jennifer Balch of the University of Colorado at Boulder has recently shown, using satellite data, that the grasses are progressively moving northwards in western North America, and having an effect on the distribution of wildfire.

Nowhere is the invasion of this grass so critical as in the iconic cactus regions of the south-western United States, often seen in Hollywood movies. In the ecological community that has evolved in this predominantly hot and dry landscape, the tall cacti (Saguaro cactus—*Carnegiea gigantea*) are isolated and not connected by any surface vegetation. If a cactus is hit by lightning then it will burn but the fire will not spread. The spread of cheatgrass into this landscape has resulted in a carpet of dry fuel, allowing the fire to spread from one cactus to another, thus threatening the whole ecosystem. It is possible that the ecosystem will be wiped out within only a few decades.

The problem of invasive grasses is not confined to the United States. In Australia, Gamba grass (*Andropogon gayanus*) was introduced from Africa, where it is natural to many areas of savanna, again as cattle feed. As we might expect, it is now causing serious problems for precisely the characteristics for which it was originally introduced: it grows very quickly, densely, and tall. These dense thickets of grass then make ideal fuel for fire, providing a fuel load up to eight times that of native grasses. The fires that occur naturally in many parts of Australia, especially in northern Australian savannas, have become hotter and flame heights are greater, and such fires risk the survival of the native vegetation, which is often dominated by eucalypts.

Eucalyptus plantations are also creating problems in some regions. In Australia eucalypt trees are part of a flammable ecosystem. They have evolved to cope with frequent fire and have developed methods of re-sprouting to survive fire. As we all probably know, these trees produce an oil that enhances their flammability. The trees grow quickly and hence produce a lot of potential fuel to burn, but regular surface fires reduce this surface fuel load. Such trees have been planted in dense stands in many parts of the world. This has become a problem especially in Portugal, where these extensive dense stands of eucalypts have been allowed to grow without sufficient fuel reduction measures. Portugal has a climate in which fires occur regularly, and over the past decade there has been a large increase in the number and size of devastating wildfires in these plantations that have claimed more than 100 lives. The intensity and size of the fires in these plantations, which surround many villages and through which access roads occur, are partly responsible for the high mortality rates. These fires in eucalypt plantations have also spread into pine forests, themselves with a considerable surface fuel load that is very flammable, making the fires very difficult to control (Box 9).

Megafires

The term 'megafire' has only come into the public consciousness in recent years. There are a large number of classifications of wildfire

Box 9. Fires in Portugal

Portugal has a typical Mediterranean-type climate and a distinct fire regime. The problem is that such a region has been shaped by human activity for centuries. In such regions fire has been a significant part of many agricultural practices. This led to a mosaic of different landscape types. In the 20th century there were wholesale changes to the system of agriculture with the development of a large-scale forestry programme. Into this fire-prone region, large areas of forest were planted, mainly of pine and then eucalypts. This plantation programme expanded through the 1950s, 1960s, and into the 1970s. Instead, then, of a patchwork of different types of agriculture, which ranged from pastoral to crop production, forest became the norm in many regions. Fire has become a particular problem in these large new forests, where people have yet to adapt to the idea of managing the forests sustainably, with fuel load reduction and, where appropriate, controlled burning. The result is that fires are not only inevitable (pines and eucalypts being very flammable), but they are intense and widespread, and have caused considerable loss of life. The fires of 2003 resulted in the burning of 10 per cent of the forests and 18 deaths. The fires in June and October 2017 resulted in a total of 111 deaths.

that reflect the frequency of fire, the type of fire, and its size. The term 'megafire' has often been used to indicate that a fire is larger and perhaps more intense than would normally be expected in a particular ecosystem. When we talk about more or less fire we need to define carefully what we mean. Is it that the number of fires is more or fewer, or that the area burned by fire is greater or less? Having more fires does not necessarily pose a problem. These fires may be small surface fires that reduce the surface fuel load. They may be accidental fires, natural fires, or prescribed burns for the purpose of fuel reduction. There has been public resistance to the use of prescribed burning, and support for other fuel reduction methods to be used instead. Logging to thin a forest may seem an obvious course, but may itself create problems, by building up extensive slash piles that increase the surface fuel load. Tree thinning may also open up the forest to light allowing new growth, adding to the surface fuel load, and also provide space for invasive grasses.

So why have there been more reported megafires, and what has caused the switch to such fires? We should note that the size of a fire is not necessarily the only issue here but also where the fire occurs. A small fire in one area may have more significant consequences than a large fire in another area. The switch to megafires can be considered an amalgam of three switches: the ignition switch, the fuel switch, and the aridity switch. In the western USA it has been shown that in 2017 these switches flipped unusually rapidly and stayed on for longer, resulting in an increased number of megafires.

While there are large numbers of natural wildfires in the western USA started by lightning strikes, it has been shown that 89 per cent of the fires were started by human activity. These burned over 19,000 sq. km of land. This included the Thomas Fire, one of the largest, and ignitions were increased in part because of the greater aridity and condition of the fuel. The nature of the fuel is another consideration. As we have seen, changes in the times of rainfall promoted the growth of fine fuels including exotic grasses and, together with the resistance to reduce fuel load, perhaps because of financial reasons and the increased development of urban centres, all these factors contributed to the availability of combustible fuel. This has led to an increase in fires that are more intense and larger. The third switch relates to aridity, and many areas of the western USA have experienced longer and hotter dry periods. This, together with the occurrence of high winds, led to a significant increase in fire risk. The combination of all these factors led to the 'big burn' of 2017.

The USA is not the only place where such events are being seen. Large wildfires were seen in many parts of the world from Siberia to Alaska through 2018 and 2019. Not all of these changes in wildfire may relate to anthropogenic climate change but all demand a response both from governments and the public. Few governments have grappled with the issue or adequately addressed the question of how to live sustainably in increasingly flammable landscapes.

We may have success in reducing the number of fires through fire prevention and suppression but a combination of changing fuel treatments (such as preventing prescribed burns), the spread of invasive grasses, and changes in climate leads to the increased potential for large, high-intensity wildfires or megafires (Figure 28). We may see fewer fires, but the areas in which these fires are burning are getting larger.

And contrary to expectation, the occurrence of a large fire does not make a future one in the region far less likely. As we have noted before, dead wood unconsumed by a previous fire provides fuel that can be added to by extensive regrowth of surface vegetation to introduce conditions for a new fire, which may be larger still. This has often been the case in California.

Since 2017 many fire fighters have noted a significant change in some wildfire behaviour. It is generally recognized that wildfires spread less quickly at night-time when temperatures fall and humidity rises. In some recent megafires, extremely rapid fire spread has occurred at night, and this phenomenon is yet to be fully explained. It may have an important impact upon fire suppression efforts. In addition, extremely rapid, unexpected movement of surface fires has been noted, such as occurred in the 2018 Camp Fire in California (but also seen for example in Portugal). Such developments present an important challenge not only to the scientific community and to fire fighters but also to the public as a whole. Such changes were also seen in the 2019–20

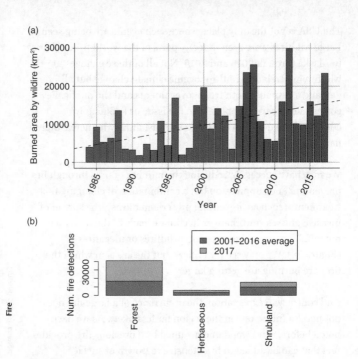

28. The megafire switch: increasing burned area trend in California wildfires

wildfires in Australia where the fires were more widespread across a range of vegetation types, were more intense and driven by strong winds; smaller fires combined to form megafires that were difficult to predict or control. In particular a large number of pyrocumulus clouds formed that not only provided a strong updraught but also created lightning that started new fires.

The future of fire in a warming world

What approaches can we make to allow us to think about fire in a world undergoing climate change? We may consider three. The first we might call experimental. Are there aspects of fire occurrence and behaviour that would be useful in assessing future fire? The second approach is through modelling. Can climate models that include fire as an element and output be developed? And can we assess the impact of fire on different types of vegetation and different regions using computer models? The third we may call the pragmatic approach: even if we do not understand or accept the different models, what can we do if fires become more severe or begin to occur where they have not before?

There are two main approaches to fire experiments. The first is at the laboratory scale. This type of approach is useful for investigating both vegetation fires and building fires, where different types of both old and new building materials can be assessed not only for flammability but also for rate of fire spread. For vegetation fires, greater understanding can be gained by investigating a range of natural fuels in different conditions and in different configurations, in terms of both ignition and spread.

While laboratory experiments can provide valuable data the larger-scale fires in a more realistic context often pose additional problems and highlight different issues. Within building fires, for example, material which had been examined and approved in laboratory tests proves sometimes to have unexpected problems when exposed to fires in a real context. Such was the case with the cladding used on the Grenfell Tower in London, where fire spread on the outside of the building was much faster and more intense than had been expected and this tragedy led to a demand to rethink the use of different cladding materials.

Experiments are also undertaken on a larger scale where, for example, buildings may be set alight and monitored to help understand how and how quickly a fire spreads. This is valuable not only for help in building design regulations or in building materials regulations but also for forensic investigations that may take place after a fire. We also need to understand the difference in how fires spread in buildings as opposed to landscape or wildfires. In building fires, a major spread direction is vertical, whereas in landscape fires the most significant direction of travel is lateral, and experience in one type of fire does not completely translate to the other fire situation, meaning that suppression training in both types of fire is essential.

In the case of wildfire too, large-scale controlled and monitored burns can be conducted through many different types of vegetation and in different conditions, including in grasslands and forests. But here also there are problems in relating these experimental burns to real situations, though there has been significant progress in this field of research since 2017. Often the fires are within relatively small areas, perhaps less than 1 sq. km. In reality, some fires change their behaviour as they develop. In addition, many of these experiments are set in conditions that would not be normal in terms of the fuel load, fuel condition, and weather. If the weather is too dry permission may not be given for a burn in case the fire escapes from the controlled area. This has happened during prescribed burns with disastrous consequences, but even in such circumstances often good data can be gathered.

All this data can be used to create models of fire at a range of scales. In the context of climate change, models have been developed that consider the future likelihood of fire in different regions in the light of different climate change scenarios. Some of these models use past data of fire to help predict fires in the future; yet others use first principles about fire occurrence and can combine global climate models with global vegetation models to indicate where fire may occur with different predicted rises in temperature. Such studies reveal that fire may become more severe in some places but less in others but that some regions without extensive experience of wildfire today may experience increasing number of fires over the next twenty to fifty years. We need to start planning now for such eventualities and we need to learn lessons from current fires. It is clear that when it comes to wildfire, you need to expect the unexpected and plan for the worst, not just hope for the best. We need more discussion about the issues raised by climate change and wildfire, involving the general public as well as fire fighters, foresters, land managers, conservationists, and policy makers.

Future expectations

Should we be positive or negative about fire in the future? In some senses we should have a positive attitude, as we are continuing to improve how we view and understand fire, allowing us to make more informed decisions about its suppression or not. Optimism may be replaced by pessimism when we consider that some of the most powerful politicians clearly misunderstand fire (as well as climate change). Our future may, however, be complicated by a number of opposing trends that may be difficult to understand for a world that demands simple solutions to complex problems.

In some areas there may indeed be less fire. The global area burned is currently declining where humans and in some cases excessive drought remove the fuel. But such trends can change rapidly so that, for example, excessive burning of the Amazon rainforest has created alarm across the world. And more severe and larger fires in warming areas around the world, which include most fire-prone forest, is another trend.

We can add to these the fact that there is more exposure of human populations to fire due to population growth in the WUI. We may add that there has been an increase in the number of ignitions due to human activity but in contrast there are also fewer accidental fires and fewer fatalities from fire as a result of advances in education and technology.

One proposed approach to reducing atmospheric carbon dioxide is the planting of extensive forests in some regions. But this needs careful consideration. If the wrong trees are planted in the wrong places then more problems may be created than they are designed to solve. The issues around fire, vegetation, and climate need much more informed discussion.

The key to managing fire lies not only in developments in research and technology but also in improving the understanding of fire among both policy makers and the wider public across the world.

We all need to think about fire—after all, where would we be without it?

References and guides

Balch, J. K., Bradley, B. A., D'Antonio, C. M., and Gomez-Dans, J. 2013. Introduced annual grass increases regional fire activity across the arid western USA (1980–2009). *Global Change Biology* 19, 173–83.

Balch, J. K., Schoennagel, T., Williams, A. P., Abatzoglou, J. T., Cattau, M. E., Nathan, P., Mietkiewicz, N. P. I., and St. Denis, L. A. 2018. Switching on the Big Burn of 2017. *Fire* 1, 17; doi:10.3390/fire1010017.

Belcher, C. M. (ed.) 2013. *Fire Phenomena in the Earth System—An Interdisciplinary Approach to Fire Science*. J. Wiley and Sons, New York.

Bond, W. J. 2016. Ancient grasslands at risk. *Science* 351 (6269), 120–2.

Bond, W. J. and Keeley, J. E. 2005. Fire as global 'herbivore': the ecology and evolution of flammable ecosystems. *Trends in Ecology and Evolution* 20, 387–94.

Bonta, M., Gosford, R., Eussen, D., Ferguson, N., Loveless, E., and Witwer, M. 2017. Intentional fire-spreading by 'firehawk' raptors in Northern Australia. *Journal of Ethnobotany* 37, 700–18.

Bowman, D. J. M. S., Balch, J. K., Artaxo, P., Bond, W. J., Carlson, J. M., Cochrane, M. A., D'Antonio, C. M., DeFries, R. S., Doyle, J. C., Harrison, S. P., Johnston, F. H., Keeley, J. E., Krawchuk, M. A., Kull, C. A., Marston, J. B., Moritz, M. A., Prentice, I. C., Roos, C. I., Scott, A. C., Swetnam, T. W., van der Werf, G. R., and Pyne, S. J. 2009. Fire in the Earth System. *Science* 324, 481–4.

Bowman, D. J. M. S., Balch, J., Artaxo, P., Bond, W. J., Cochrane, M. A., D'Antonio, C. M., DeFries, R., Johnston, F. H., Keeley, J. E.,

Krawchuk, M. A., Kull, C. A., Mack, M., Moritz, M. A., Pyne, S. J., Roos, C. I., Scott, A. C., Sodhi, N. S., and Swetnam, T. W. 2011. The human dimension of fire regimes on Earth. *Journal of Biogeography* 38, 2223–36.

Cerdà, A. and Robichaud, P. (eds). 2009. *Fire Effects on Soils and Restoration Strategies*. Science Publishers Inc. Enfield, NH.

Cheney, P. and Sullivan, A. 2008. *Grassfires: Fuel, Weather and Fire Behaviour*. CSIRO, Melbourne.

Clark, J. S. et al. 1991. Sediment records of biomass burning and global change, pp. 347–65. *NATO ASI Series I*, Vol. 51. Springer-Verlag.

Cochrane, M. A. (ed.) 2009. *Tropical Fire Ecology: Climate Change, Land Use and Ecosystem Dynamics*, pp. 24–62. Springer, Berlin.

Cohen, J. 2008. The wildland–urban interface fire problem: a consequence of the fire exclusion paradigm. *Forest History Today* Fall issue, 20–6.

Davis, K. P. (ed.) 1959. *Forest Fire: Control and Use*. McGraw-Hill, New York.

Flannigan, M. D., Stocks, B. J., and Wotton, B. M. 2000. Climate change and forest fires. *The Science of the Total Environment* 262, 221–9.

Forestry Commission 2014. *Building Wildfire Resilience into Forest Management Planning*. Forestry Commission Practice Guide Forestry Commission, Edinburgh. <https://www.forestresearch. gov.uk/research/building-wildfire-resilience-into-forest-management-planning/>.

Graham, R. T. (ed.) 2003. Hayman Fire Case Study. Gen. Tech. Rep. RMRS-GTR-114. Ogden, Ut.: US Department of Agriculture, Forest Service, Rocky Mountain Research Station. 396 pp.

He, T., Lamont, B. B., and Downes, K. S. 2011. *Banksia* born to burn. *New Phytologist* 191, 184–96.

He, T., Pausas, J. G., Belcher, C. M., Schwilk, D. W., and Lamont, B. B. 2012. Fire-adapted traits of *Pinus* arose in the fiery Cretaceous. *New Phytologist* 194, 751–9.

Johnson, B. 1984. *The Great Fire of Borneo: Report of a Visit to Kalimantan-Timur a Year Later, May 1984*. World Wildlife Fund, Godalming, 24 pp.

Johnston, F. H., Henderson, S. B., Chen, Y., Randerson, J. T., Marlier, M., DeFries, R. S., Kinney, P., Bowman, D. M. S., and Brauer, M. 2012. Estimated global mortality attributable to smoke from landscape fires. *Environmental Health Perspectives* 120, 695–701.

Keeley, J. E., Bond, W. J., Bradstock, R. A., Pausas, J. G., and Rundel, P. W. 2012. *Fire in Mediterranean Climate Ecosystems: Ecology, Evolution and Management*. Cambridge University Press, Cambridge.

Krawchuk, M. A., Moritz, M. A., Parisien, M.-A., Van Dorn, J., and Hayhoe, K. 2009. Global pyrogeography: the current and future distribution of wildfire. *PloS One* 4(4), e5102.

McParland, L. C., Collinson, M. E., Scott, A. C., and Campbell, G. 2009a. The use of reflectance for the interpretation of natural and anthropogenic charcoal assemblages. *Archaeological and Anthropological Sciences* 1, 249–61.

McParland, L. C., Hazell, Z., Campbell, G., Collinson, M. E., and Scott, A. C. 2009b. How the Romans got themselves into hot water: temperatures and fuel types of a Roman hypocaust fire. *Environmental Archaeology* 14, 172–9.

Marlon, J. R., Bartlein, P. J., Walsh, M. K., Harrison, S. P., Brown, K. J., Edwards, M. E., Higuera, P. E., Power, M. J., Anderson, R. S., Briles, C., Brunelle, A., Carcaillet, C., Daniels, M., Hu, F. S., Lavoie, M., Long, C., Minckley, T., Richard, P. J. H., Scott, A. C., Shafer, D. S., Tinner, W., Umbanhowar, C. E,. Jr., and Whitlock, C. 2009. Wildfire responses to abrupt climate change in North America. *Proceedings of the National Academy of Sciences, USA* 106, 2519–24.

Mooney, H. A., Bonnicksen, T. H., Christensen, N. L., Lotan, J. E., and Reiners, W. A. (eds). 1981. Fire regimes and ecosystem properties, pp. 401–20. USDA For. Serv. Gen. Tech. Rep., WO-26.

Nature Geoscience special issue set of papers 2019: <https://www.nature.com/collections/jchbhhagcb>.

Pausas, J. G. 2015. Alternative fire-driven vegetation states. *Journal of Vegetation Science* 26, 4–6.

Pausas, J. G. 2019. Generalized fire response strategies in plants and animals. *Oikos* 128, 147–53, 2019 doi: 10.1111/oik.05907.

Pausas, J. G. and Parr, C. L. 2018. Towards an understanding of the evolutionary role of fire in animals. *Evol. Ecol.* 32, 113–25. <https://doi.org/10.1007/s10682-018-9927-6>.

Pausas, J. G. and Keeley, J. E. 2009. A burning story: the role of fire in the history of life. *Bioscience* 59, 593–601.

Pausas, J. G. and Keeley, J. E. 2014. Abrupt climate-independent fire regime changes. *Ecosystems* 17, 1109–20 doi: 10.1007/s10021-014-9773-5.

Pierce, J. L., Meyer, G. A., and Jull, A. J. T. 2004. Fire-induced erosion and millennial-scale climate change in northern ponderosa pine forests. *Nature* 432, 87–90.

Pyne, S. J., Andrews, P. L., and Laven, R. D. 1996. *Introduction to Wildland Fire*. J. Wiley and Sons, New York.

Scott, A. C. and Damblon, F. (eds) 2010. Charcoal and its use in palaeoenvironmental analysis. *Palaeogeography, Palaeoclimatology, Palaeoecology* 291, 1–165.

Scott, A. C., Moore, J., and Brayshay, B. (eds) 2000. Fire and the palaeoenvironment. *Palaeogeography, Palaeoclimatology, Palaeoecology* 164, 1–412.

Smith, A. M. S., Kolden, C. A., and Bowman, D. J. M. S. 2018. Biomimicry can help humans to coexist sustainably with fire. *Nature Ecology & Evolution* 2, 1827–9.

Sugihara, N. G., Van Wagtendonk, J. W., Shaffer, K. E., Fites, Kaufman, J., and Thode, A. E. (eds) 2018. *Fire in California's Ecosystems*. 2nd edition. 568 pp. University of California Press, Berkeley.

Further reading

Alexander, M. E., Mutch, R. W., Davis, K. M., Bucks, C. M. 2017.
Wildland fires: dangers and survival. Pp. 276–318 in
P. S. Auerbach, ed., *Auerbach's Wilderness Medicine*, Volume 1.
7th edition. Elsevier, Philadelphia.

Billings Gazette. 1995. *Yellowstone on Fire*. Revised edition. Billings,
Montana. 128 pp.

Burton, F. D. 2009. *Fire: The Spark that Ignited Human Evolution*.
University of New Mexico Press, Albuquerque. 231 pp.

Castellnou, M., Alfaro, L., Miralles, M., et al. 2019. Learning to fight a
new kind of fire. *Wildfire Magazine* 28 (4), 26–34.

Connors, P. 2011. *Fire Season: Field Notes from a Wilderness Lookout*.
McMillan, Basingstoke. 246 pp.

Cottrell, W. H., Jr 2004. *The Book of Fire*. 2nd edition. Mountain
Press, Missoula, Mont. 74 pp.

Drysdale, D. 2011. *An Introduction to Fire Dynamics*. 3rd edition.
Wiley, Chichester. 551 pp.

Hansen, C. and Griffiths, T. 2009. *Living with Fire: People, Nature
and History in Streels Creek*. CSIRO Publishing, Collingwood,
Victoria. 190 pp.

Hawley, J., Hurley, G., and Sackett, S. 2017. *Into the Fire: The Fight to
Save Fort McMurray*. McClelland & Stewart, Toronto. 160 pp.

Johnson, B. 1984. *The Great Fire of Borneo: Report of a Visit to
Kalimantan-Timur a Year Later, May 1984*. World Wildlife Fund,
Godalming. 24 pp.

Kennedy, R. G. 2006. *Wildfire and Americans*. Hill and Wang,
New York. 332 pp.

Kodas, M. 2017. *Megafire: The Race to Extinguish a Deadly Epidemic
of Flame*. Houghton Mifflin Harcourt, New York. 395 pp.

Parliamentary Office of Science and Technology 2019. POST note 603. Climate change and UK wildfire. <https://researchbriefings. parliament.uk/ResearchBriefing/Summary/POST-PN-0603>.

Pereira, J. S., Pereira, J. M. C., Rego, F. C., Silva, J. N., and Silva, T. P. (eds) 2006. *Incêndios Florestais em Portugal*. ISA Press, Lisbon.

Pyne, S. J. 1997. *Vestal Fire: An Environmental History, Told through Fire, of Europe and of Europe's Encounter with the World*. University of Washington Press, Washington.

Pyne, S. J. 2001. *Year of the Fires: The Story of the Great Fires of 2010*. Penguin Books, New York. 322 pp.

Pyne, S. J. 2012. *Fire: Nature and Culture*. Reaktion Books, London.

Pyne, S. J. 2019. *Fire: A Brief History*. 2nd edition. University of Washington Press, Washington. 216 pp.

Quintiere, J. G. 1998. *Principles of Fire Behavior*. Delmar Publishers, Albany, NY. 258 pp.

Quintiere, J. G. 2006. *Fundamentals of Fire Phenomena*. J. Wiley, Chichester. 439 pp.

Rossotti, H. 1993. *Fire: Technology, Symbolism, Ecology, Science, Hazard*. Oxford University Press, Oxford. 288 pp.

Scott, A. C. 2018. *Burning Planet. The Story of Fire through Time*. Oxford University Press. 224 pp.

Scott, A. C., Bowman, D. J. M. S., Bond, W. J., Pyne, S. J., and Alexander, M. E. 2014. *Fire on Earth: An Introduction*. J. Wiley and Sons., New York. 413 pp.

Scott, A. C., Chaloner, W. G., Belcher, C. M., and Roos, C. I. (eds) 2016. The interaction of fire and mankind: *Phil. Trans. R. Soc. B*. 371 (1696), 252 pp.

Thomas, P. A. and McAlpine, R. S. 2010. *Fire in the Forest*. Cambridge University Press, Cambridge. 225 pp.

Wallace, L. L. (ed.) 2004. *After the Fires: The Ecology of Change in Yellowstone National Park*. Yale University Press, New Haven. 390 pp.

Wrangham, R. 2009. *Catching Fire: How Cooking Made us Human*. Profile Books, London. 309 pp.

Publisher's acknowledgements

We are grateful for permission to include the following copyright material in this book.

Box 7, based on information courtesy of University of Nevada Cooperative Extension Service.

The publisher and author have made every effort to trace and contact all copyright holders before publication. If notified, the publisher will be pleased to rectify any errors or omissions at the earliest opportunity.

Publisher's acknowledgement

Index

For the benefit of digital users, indexed terms that span two pages (e.g., 52–53) may, on occasion, appear on only one of those pages.

EPIDEMIOLOGY
A Very Short Introduction
Rodolfo Saracci

Epidemiology has had an impact on many areas of medicine; and lung cancer, to the origin and spread of new epidemics. and lung cancer, to the origin and spread of new epidemics. However, it is often poorly understood, largely due to misrepresentations in the media. In this *Very Short Introduction* Rodolfo Saracci dispels some of the myths surrounding the study of epidemiology. He provides a general explanation of the principles behind clinical trials, and explains the nature of basic statistics concerning disease. He also looks at the ethical and political issues related to obtaining and using information concerning patients, and trials involving placebos.

Geopolitics

A Very Short Introduction
Klaus Dodds

In certain places such as Iraq or Lebanon, moving a few feet either side of a territorial boundary can be a matter of life or death, dramatically highlighting the connections between place and politics. For a country's location and size as well as its sovereignty and resources all affect how the people that live there understand and interact with the wider world. Using wide-ranging examples, from historical maps to James Bond films and the rhetoric of political leaders like Churchill and George W. Bush, this Very Short Introduction shows why, for a full understanding of contemporary global politics, it is not just smart - it is essential - to be geopolitical.

'Engrossing study of a complex topic.'

Mick Herron, Geographical.

MODERN CHINA
A Very Short Introduction
Rana Mitter

China today is never out of the news: from human rights controversies and the continued legacy of Tiananmen Square, to global coverage of the Beijing Olympics, and the Chinese 'economic miracle'. It seems a country of contradictions: a peasant society with some of the world's most futuristic cities, heir to an ancient civilization that is still trying to find a modern identity. This *Very Short Introduction* offers the reader with no previous knowledge of China a variety of ways to understand the world's most populous nation, giving a short, integrated picture of modern Chinese society, culture, economy, politics and art.

'A brilliant essay.'

Timothy Garton, TLS